endorsed for
edexcel ::::

REVISE EDEXCEL GCSE
Statistics

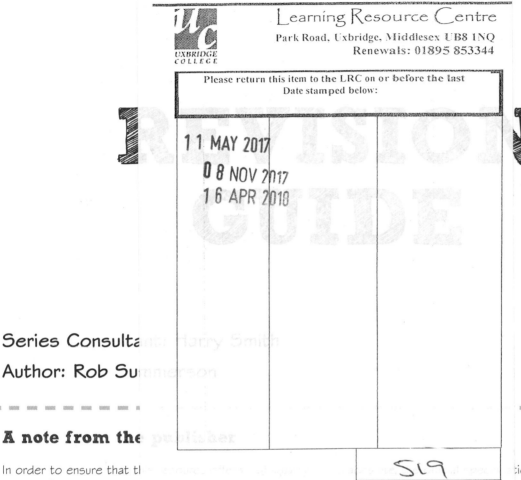

Series Consultant: Harry Smith

Author: Rob Summerson

A note from the publisher

In order to ensure that th... ...tion and associated
support for the associat... ...are the only authoritative
been through a review pr... ...always be referred to
This process confirms that this resource fully covers the
teaching and learning content of the specification or part
of a specification at which it is aimed. It also confirms
that it demonstrates an appropriate balance between
the development of subject skills, knowledge and
understanding, in addition to preparation for assessment.

Endorsement does not cover any guidance on
assessment activities or processes (e.g. practice
questions or advice on how to answer assessment
questions), included in the resource nor does it
prescribe any particular approach to the teaching or
delivery of a related course.

While the publishers have made every attempt to ensure
that advice on the qualification and its assessment

Pearson examiners have not contributed to any sections
in this resource relevant to examination papers for which
they have responsibility.

Examiners will not use endorsed resources as a source
of material for any assessment set by Pearson.

Endorsement of a resource does not mean that the
resource is required to achieve this Pearson qualification,
nor does it mean that it is the only suitable material
available to support the qualification, and any resource
lists produced by the awarding bod...
and other appropriate res...

**For the full range of Pearson revision titles across GCSE,
AS/A Level and BTEC visit:**
www.pearsonschools.co.uk/revise

ALWAYS LEARNING

PEARSON

Contents

1-to-1 page match with the Statistics Revision Workbook ISBN 9781292098289

. . . .

A small bit of small print

Edexcel publishes Sample Assessment Material and the Specification on its website. This is the official content and this book should be used in conjunction with it. The questions in *Now try this* have been written to help you practise every topic in the book. Remember: the real exam questions may not look like this.

Types of data

People use statistics to analyse data. This page summarises what you need to know about data. There are a lot of definitions but you could be asked about any of them in the exam.

Primary data

Primary data is information that you collect yourself.

You could do an experiment, carry out a survey or use a questionnaire to collect primary data.

Secondary data

Secondary data comes from published sources, such as newspapers, books or the internet.

You could take information from a table in a magazine to collect secondary data.

Categorical data

Categorical data can be sorted into groups.

Categorical data

Can be counted or measured.

Qualitative   Quantitative

e.g. gender

Numerical (discrete)

e.g. the number of children in a family

Measurements (continuous)

e.g. the weight of a child

Ranked (discrete)

e.g. the position in a race

Worked example

tier F

Here is a list of statistical words:

    qualitative   quantitative   categorical
        ranked   measurement

Choose the word from the list that best describes the data below. If none of the words is suitable then write 'none'.

(a) Number of cars in a car park   (1)
Quantitative

(b) Shoe size   (1)
Quantitative

(c) Position in a class test   (1)
Ranked

(d) Style of a painting   (1)
None

(a) and (b) are quantitative since it is a number in each case.

Positions are the same as ranks.

Now try this

tier F

Look at the diagram if you get stuck.

1 Here are descriptions of four sets of data.
  A Football scores taken from a magazine.
  B Heights of plants you grew under different light conditions.
  C Types of adverts that you recorded on television one night.
  D Positions of football teams in a league taken from the internet.

Which of these sets of data are
(a) primary data
(b) secondary data
(c) qualitative data
(c) quantitative continuous data
(d) quantitative discrete data?   (2)

Measurements and variable

Measurements

Measurements can be taken only to a certain level of accuracy (e.g. 1 mm) so measured quantitative data is always approximate.

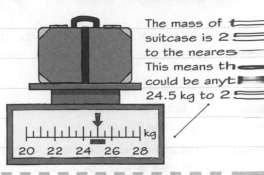

The mass of t
suitcase is 2
to the neares
This means th
could be anyt
24.5 kg to 2

Variables

A **variable** in statistics is something that can vary which is being observed, e.g. eye colour varies from person to person, height varies from plant to plant.

Bivariate data are pairs of related variables. These are usually quantitative but can be qualitative or even one of each.

Bivariate data can be shown in a table or in a graph.

Name	Age (years)	Length of foot (cm)
Bill	4	16
Jo	6	18
Sanjay	7	17
Chen	12	22

Explanatory and response variables

When a scientist does an experiment, the variable the scientist has control over is the **explanatory** (or **independent**) variable.

The measured outcomes are described the **response** (or **dependent**) variable.

In the example, the two variables are age and length of foot.

Age is the explanatory variable.

Length of foot is the response variable.

The explanatory variable is on the horizontal axis.

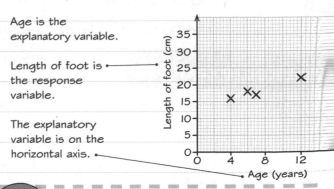

Worked example

tier F

In each case state whether the bivariate data are quantitative or qualitative and which is the explanatory variable.
(a) The age of a child and his or her height. (2)
Both age and height are quantitative.
The explanatory variable is age.
(b) The mass of a person and their gender. (2)
Mass is quantitative, gender is qualitative.
Gender is the explanatory variable.

EXAM ALERT!

Students have struggled with the idea of explanatory variable. 'Age' is the explanator (or independent) variable because height depends on age and not the other way round.

Students have struggled with exam questions similar to this – **be prepared!**

Results**Plus**

Now try this

tier F&H

1 The area of a lawn is 24 m² correct to the nearest m². Give
   (a) the least value the measurement could be
   (b) the greatest value the measurement could be. (2)
2 In each case state whether the bivariate data are quantitative or qualitative and which is the explanatory variable.
   (a) The total number of eggs a hen has laid in its life and its age.
   (b) The price of a second-hand car and its age. (2)

Sampling frames, pre-tests and pilots

Technical words and phrases

You need to know these definitions:

A **population** is everything or everybody that could possibly be involved in an investigation, e.g. students in a school, all the people who use the local gym.

A **census** gathers data from the whole population.

A **sample** gathers data from some of the population.

A **sampling frame** is a list of all the members of the population from which the sample will be taken.

A **pilot** is a small sample analysed first before any large-scale samples.

A **pre-test** is a pilot where questions for a questionnaire are usually tried out.

Census vs sample

Here is a population.

A census would gather information from **everyone**.

All the members of the population can be numbered to form a **sampling frame**.

A **random sample** is shown in red.

A census collects more information than a sample but takes a lot longer and is a lot more expensive.

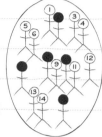

Worked example

tier F&H

(a) Explain what is meant by a 'random sample'. **(1)**

A random sample is one in which every member of the population has the same chance of being selected.

(b) Amina is going to take a random sample of 200 houses from the 4000 houses in her town. Explain how Amina could carry out a random sample. **(2)**

Write a sampling frame by listing the street names and the house numbers in order.

Give each house a different number from 1 to 4000.

Generate 200 random numbers between 1 and 4000.

Use these numbers to pick the houses in her sample.

284   553   341   034   697   451   107

> You need to learn this definition. Don't use 'random' or 'no pattern' in your answer.

> Random numbers can be found on the internet or generated on your calculator. Use the (ran#) button and ignore the decimal point.

Good and poor samples

Good samples	Poor samples
are as large as possible	are too small
are unbiased	are biased – they unfairly favour one set of values
have a suitable sampling frame	have a poor sampling frame, e.g. out of date, people missing, people counted twice, names on the list that should not be there

---

Now try this

tier F

1 Jim wants to find out how many of the 250 students in his year bring a mobile phone to school. He decides to ask 10 of his friends.

(a) Write down two reasons why this is a not a good sample.

(b) Explain how Jim could take a better sample. **(3)**

Experiments and hypotheses

Data can be produced when you carry out an experiment on a sample. You can then use results from the sample to make predictions about the population.

Experiments

A **hypothesis** is a statement of belief about some aspect of a population.

You can carry out **experiments** or make observations to see if your hypothesis is supported by the data you collect.

A **control** in an experiment is designed to check the hypothesis. The experimenter compares their results against this standard.

A **sample** can be used to make predictions about the population or comparisons between parts of a population.

The flow chart on the right describes a **statistical experiment**. Box 1 states the **hypothesis** being tested and box 3 describes the **control**.

 1 Adverts are more likely to be targeted at stay-at-home mums during daytime TV than during evening TV.

↓

 2 Collect information about the numbers of daytime TV adverts targeted at stay-at-home mums.

↓

3 Collect information about the numbers of adverts in the evening targeted at stay-at-home mums.

↓

 4 Compare the data from 2 and 3.

Worked example

 tier **F**

Jenny makes a hypothesis that seeds need water to germinate and grow. She plants 10 seeds and waters them. They all grow so she says her hypothesis was correct.
Explain what is wrong with Jenny's experiment. **(2)**

She needs to try to grow some seeds that are not watered. This is her control. She should then compare the growth of the watered seeds and the unwatered seeds.

It's a good idea to show you understand by giving an example of how Jenny can improve her experiment.

Worked example

tier **F&**

A random sample of 200 houses was taken from the 4000 houses in a town. The number of houses in the sample that had solar panels was 12.
(a) Work out an estimate for the number of houses in the town with solar panels. **(2)**

Proportion of houses with solar panels in the sample is $\frac{12}{200}$

An estimate for the number of houses with solar panels in the town is

$\frac{12}{200} \times 4000 = 240$

(b) How could a better estimate be obtained? **(1)**

By taking a bigger sample.

You can use proportion for part (a). Work out the proportion of houses with solar panels in the sample, then multiply by the size of the population. For part (b), the bigger the sample, the more accurate the estimate will be.

Now try this

 Don't forget the control.

tier **F&H**

1 Giles wants to know whether spraying plants with detergent will get rid of pests.
Design an experiment for Giles. **(2)**

2 In a sample of 320 young people living in a town, 26 said they could not afford to go to college. 14 000 young people live in the town.
Work out an estimate for the number of young people in the town who would say they could not afford college. **(2)**

3 In a sample of 50 students from a school, 38 had a mobile phone with them.
Work out an estimate for the percentage of students at the school who had a mobile phone with them. **(2)**

Stratified sampling

Stratified sampling is a sampling method which can be used when the population is known to be split into distinct **groups**. The advantage over simple random sampling is that you can be certain that all the groups, even the small ones, are represented in the sample. This may not be true for simple random sampling.

Strata

A **stratum** is a group in the population. In a stratified sample, the experimenter ensures that the relative sizes of the groups in the sample are the same as their relative sizes in the whole population.

There are twice as many boys as girls in this population.

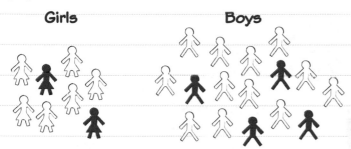

Girls   Boys

So you need twice as many boys as girls in the stratified sample (shown in red).

Worked example

This table shows the numbers of males and females who belong to a gym.

Males	Females	Total
120	80	200

A sample of 40 gym members stratified by gender is to be taken. Work out the number of males in the sample. **(2)**

Proportion of males in population
$$= \frac{120}{200} = 0.6$$
Number of males in sample
$$= 0.6 \times 40 = 24$$

The proportion of males in the sample must be the same as the proportion of males in the population. In order to select this sample the sampling frame must contain data about the gender of each member of the population.

Worked example

A primary school has four classes in year 6.

Class	A	B	C	D
Number of children	34	28	18	10

The head teacher wants to take a sample of 30 children stratified by class.
(a) Work out the number of children from class C who should be in the sample. **(2)**

There are 34 + 28 + 18 + 10 = 90 children in Year 6.
The proportion of class C in the population is $\frac{18}{90}$
The number from class C in the sample should be
$$\frac{18}{90} \times 30 = 6$$

(b) Work out the number of children from class B who should be in the sample. **(1)**
For class B, $\frac{28}{90} \times 30 = 9.333...$
So 9 children from class B should be in the sample.

If the total size of the population is not given then you need to work it out.

You can use either the fraction $\frac{18}{90}$ or the decimal 0.2 as the proportion.

You cannot have a decimal number of children in the sample, so round to the nearest integer.

Now try this

1 There are 520 young people in a youth club. 280 are boys.
   The youth club leader wants to take a stratified sample of 50 young people.
   Work out how many boys should be in the sample. **(2)**

Further stratified sampling

Stratified sampling using two-way tables

You can use each cell in a two-way table as a stratum.

You need to ensure that the proportion of the stratum in the sample is the same as its proportion in the population.

The table shows the numbers of female and male car and HGV drivers.

One stratum (one of the cells) is that of female car drivers.

	Female	Male
Car drivers	1543	1878
HGV drivers	238	3789

The proportion of this stratum in the population is
$$\frac{1543}{1543 + 1878 + 238 + 3789} = \frac{1543}{7448}$$
So the number of female car drivers in the sample should be $\frac{1543}{7448} \times$ the size of the total sample.

Worked example

 tier **H**   Aiming higher

Here are the numbers of people who use a library.

	Female	Male
Under 25	1543	878
25 and over	5060	3789

Jan wants to take a sample of 120 library users stratified by age and gender.

	Female	Male
Under 25	16	9
25 and over	54 55	40

Complete this table to show how many people in each of the four groups should be in the sample. **(4)**

Females under 25:
Population size is
1543 + 5060 + 878 + 3789 = 11270

Proportion of under-25 females = $\frac{1543}{11270}$

$\frac{1543}{11270} \times 120 = 16.42$, rounded to 16

Females 25 and over:
Population size is the same = 11270

Proportion of 25+ females = $\frac{5060}{11270}$

$\frac{5060}{11270} \times 120 = 53.87$, rounded to 54

Follow these steps:

1. Work out the total population size.
2. Calculate the proportion for each stratum.
3. Multiply by the population size for each stratum.
4. Round to the nearest whole number.

You need to check that your total sample is the correct size.
16 + 9 + 54 + 40 = 119 so add 1 to the stratum of females 25 and over.

Golden rule

If your sample size is too small you need to add 1 to the largest stratum.

Now try this

 tier **H**   Aiming higher

1 Here are the numbers of patients attending a surgery.

	Male	Female
Adult	48	64
Child	28	30

The doctor wants to take a sample of 50 patients stratified by age and gender. Work out the number of male adults there should be in the sample. **(3)**

Further sampling methods

Here are three other methods of sampling you need to know: **cluster** sampling, **quota** sampling and **systematic** sampling.

1 Cluster sampling

Cluster sampling can be used when the population is in groups. A random sample of these groups is selected and all items in the selected groups are included in the sample.

> Cluster sampling is a practical way to bring down the costs of sampling.

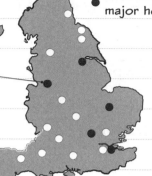

A map showing the locations of major hospitals in England.

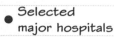
Selected major hospitals

A sample of surgeons has to be taken. Cluster sampling involves selecting major hospitals at random and then surveying all the surgeons at each selected hospital.

2 Quota sampling

Quota sampling involves splitting the population into groups with certain characteristics (e.g. age, gender) and selecting a given number from each group. For example, a market researcher might ask 10 adults and 10 children about their reaction to the 2015 GCSE results.

> Quota sampling is often used in market research and polling.

3 Systematic sampling

In systematic sampling items are selected from the population at regular intervals either in time or in space. For example, every 5th car that passes a location or every 3rd house on a street.

Worked example

tier H

Chocolate bars are made on a production line. Every 200th bar is taken away and tested for quality.
(a) Name the type of sampling being used. **(1)**
Systematic sampling
The company makes 8000 bars each day.
It decides to sample 2% of the bars.
(b) Describe how the company could take a systematic sample of 2% of the bars. **(2)**
2% of 8000 = 160
They should sample 160 bars (every 50th bar).

Sometimes a question might ask you to name a disadvantage of a sampling system. Here, there might be a hidden problem in the machinery which only occurs with odd-numbered bars and so never appears on every 200th bar.

To find 2% of 8000 work out $\frac{2}{100} \times 8000$

Now try this

tier H

Remember to answer the question in context. So here, make sure you refer to the house numbers on each side of the street.

1 A street contains 160 houses. There are 80 houses on each side of the street, arranged in 20 blocks of 4 houses each. A researcher wants to take a systematic sample of 40 houses from the street.
   (a) Describe a simple way in which he could do this.
   (b) Describe one disadvantage of this.
   (c) Describe how to re-design the systematic sample to reduce the impact of the disadvantage you have described in part (b). **(4)**

2 About 70% of expensive women's perfume is bought by women for themselves and about 30% is bought by men for women.
A marketing company wants to test out people's opinion of a new perfume. It has enough money to interview 40 people and must get the results quickly.
   (a) Explain why quota sampling may be a suitable method.
   (b) Describe how this could be carried out. **(3)**

Sampling overview

Different methods of data collection have different advantages and disadvantages. This table summarises the different methods.

Method	Advantages	Disadvantages
Census	Collects **all** the data	Slow, so the population may change before the census is finished Expensive
Simple random	Unbiased, easy to understand	Can take a long time Does not give results as representative as stratified sampling
Stratified	Unbiased, most accurate	Harder to analyse data Need to know about the strata in advance
Cluster	Quicker than random	Leads to less reliable data than a random sample
Systematic	Can be more convenient	Not a random sample, so less representative
Quota	Quicker and cheaper	Not a random sample, so less representative

Worked example

tier **H**   Aiming higher

(a) Describe the difference between a stratified sample and a cluster sample. **(1)**

Stratified sample – the strata are selected in advance and a random sample is taken from each stratum.

Cluster sample – the clusters are selected at random and then all the items in the selected clusters are included.

> You are expected to know what the different types of sampling are and to be able to describe them.

A city has 250 dental surgeries employing over 600 dentists in total. A researcher wants to carry out face-to-face interviews with a sample of 70 dentists.

(b) Explain why a cluster sample would be a suitable way of carrying out the interviews. **(1)**

The surgeries are likely to be geographically spread out, so it's more efficient to interview all the dentists at a small number of surgeries.

> You must answer in context and not just give a general definition of cluster sampling. Since the surgeries are in a city, a great deal of travelling would be required.

(c) Describe what the sampling frame would be in this case. **(1)**

A list of all the dental surgeries in the city, written in alphabetical order or postcode order. This would make it easy to take a random sample.

> You must answer in context and not just give a general definition of a sampling frame. Here you must refer to a list of surgeries and how they are ordered.

Now try this

tier **H**   Aiming higher

1 A researcher is studying the amount of pollution from cars. He takes a sample of 10 cars every four hours.
  (a) Name the type of sampling. **(1)**
  (b) Describe one possible defect of this sampling method. **(1)**

2 (a) When should you use a stratified sample? **(1)**
  A polling company wants to predict the result of an election in a town. Nationally it knows that older voters will tend to vote for the government party and younger voters for the opposition. There are 30 000 older voters and 10 000 younger voters in the town.
  (b) Explain how the polling company should carry out its poll. **(2)**

Data capture sheets

Collecting data is very important in statistics. You need to know how to do it properly and how to criticise a method of data collection.

Data capture sheets

You can collect data using a data capture sheet.

The data capture sheet should have three columns:

1   **subject** – the names of what you are collecting information on

2   **tally** – marks ⌗ written as you count

3   **frequency** – the total of the row of tally marks.

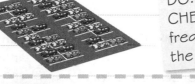

The first column lists all the car colours seen.

The tally column shows the tallies. Group them in 5s.

Colour of car	Tally	Frequency
Blue	||	2
Green	⌗ ||	7
Red	⌗ ⌗ |||	13
Yellow	⌗ ⌗ ||	12

The frequency column shows how many cars of each colour were seen.

DO: Label the three columns.
DO: Use a ruler to draw your table.
CHECK: Add up the numbers in the frequency column and check against the total number of cars.

Worked example

tier F

Here is a list of the 20 trees in a wood.

Ash   Oak   Hazel   Ash   Beech
Hazel   Ash   Hazel   Ash   Beech
Hazel   Ash   Ash   Hazel   Hazel
Beech   Beech   Hazel   Oak   Hazel

Use a data capture sheet to record this information. **(3)**

Type of tree	Tally	Frequency
Ash	⌗ |	6
Oak	||	2
Hazel	⌗ |||	8
Beech	||||	4

Check it!
6 + 2 + 8 + 4 = 20 ✓

To design a data capture sheet:

1. Look through to see how many different types of trees there are.
2. Draw a three-column table with headings for Type of tree, Tally and Frequency.
3. Write the names of the different types of trees in the first column.
4. Use the Tally column to record the number of each type of tree. Cross each tree off the list as you tally it.
5. Add up the tallies for each row and write the totals in the Frequency column.

EXAM ALERT!

Some students forget to put the heading for the first column. You should always have three headings for any data capture sheet.

Students have struggled with exam questions similar to this – **be prepared!**

ResultsPlus

Now try this

tier F

1 Jessie wants to find out the number of men, the number of women, the number of boys and the number of girls using a sports centre one night.
Design a data capture sheet that she could use. **(3)**

2 Here is a list of some precious stones:
Ruby Diamond Emerald Ruby Sapphire
Ruby Emerald Ruby Diamond Emerald
Sapphire Sapphire Ruby Emerald Ruby
Diamond Diamond Sapphire Emerald
Emerald Ruby Diamond Ruby Sapphire
Record this information in a data capture sheet. **(3)**

Interviews and questionnaires

Interviews vs questionnaires

Interviews are usually carried out person to person, with the interviewer recording the response
Questionnaires are usually posted online or sent to people for them to respond to and return

	Advantages	Disadvantages
Interviews	Interviewer can explain complex questions	Interviewer may be biased
	Interviewer can follow up on unclear responses	Can be costly
Questionnaires	Much cheaper to do	Can be inflexible
	Each person answering the question is treated in the same way	People may misunderstand some questions

Types of questions

Open questions allow a wide variety of responses,

e.g. "What do you think about programmes on TV?"

Closed questions restrict the possible replies given,

e.g. "Are you over 18 years old?"

Usually, closed questions have a set of response boxes for the respondent to tick.

Response boxes should:

- cover all possible answers
- not contain overlapping answers
- be clearly labelled.

Avoid **leading questions** which might **lead** the respondent towards the answer that you want or expect,

e.g. "How inhumane do you find the gassing of badgers?"

Advantages and disadvantages

Open questions:
- ☺ can produce unexpected answers
- ☹ are time-consuming to analyse.

Closed questions:
- ☺ clarify what type of answer is required
- ☺ are easier to analyse
- ☹ can be inflexible
- ☹ can miss out unexpected responses.

Worked example

tier F&H

A manager of a company wants to find out what his employees feel about the pay they get.
He considers interviewing some of his employees.
(a) Write down one reason why this may not be a good method. **(1)**

His employees may not give him an honest answer.
He also considers a questionnaire.
One question on the questionnaire could be, "How can we make the already very good pay structure in the company even better?"
(b) Give two reasons why this is not a good question. **(2)**
1. It's a leading question.
2. It's an open question so would be hard to analyse.

You need to think about the context when answering a question like this. The employees may not give a frank opinion because he is their boss. You could also answer that the manager may be biased as an interviewer.

Leading questions are phrased in such a way that they may influence the answer given.
You can see this is an open question in that it allows many different responses.

Now try this

tier F&H

1 A doctor wants to find out whether or not people think it is easy to get an appointment. She could either interview people or use a questionnaire. Give one advantage and one disadvantage for each. **(4)**

Questionnaires

Question design

A student wants to find out how often drivers use a road.

Here is a good question: ✓ The question has a specific **time frame** ('last week') and allows a clear response.

Here is a poor question: ✗ The question does not refer to a specific period of time.

How many times have you driven on this road in the last week?

☐ Never   ☐ 1–3   ☐ 4–6   ☐ 7–9   ☐ 10 or more

How many times have you driven on this road?

☐ 1–3   ☐ 3–6   ☐ 7–8   ☐ 8+

✓ There are at least three response boxes

✓ Values in the boxes do not overlap.
✓ The boxes cover all possible responses.

✗ '0' or 'Never' is missing.

✗ Most labels are clear but there is overlap between boxes at '3'.

✗ This label is ambiguous.

Worked example
tier F&H

Akbar wants to know how far people will travel to buy organic food. He asks people this question on a questionnaire: How far do you travel to buy organic food?

☐ 1–2 km   ☐ 3–5 km   ☐ 6+ km

(a) Write down three things that are wrong with this question. **(3)**
1. No time frame.
2. No distance less than 1 km in the response boxes.
3. Gaps in the distances (e.g. 2.5 can't be placed).

(b) Write down one thing that would improve the question. **(1)**
You should have a time frame (e.g. The last time you bought organic food, how far did you travel?)

When asked to criticise a question on a questionnaire ask these questions:
- Is there a time frame?
- Are the response boxes non-overlapping?
- Do the response boxes cover all possible responses?

Golden rules

Remember these golden rules when you are designing questions for questionnaires:

✓ Make questions clear and closed.
✓ Avoid open questions.
✓ Don't ask leading questions.
✓ Have response boxes which are unambiguous.
✓ Have response boxes which cover all possible replies and don't overlap.

Now try this
tier F&H

1 Mr Brown owns a café. He wants to find out what people think of the service in the café. He uses this question in a questionnaire:
What do you think of the service in the café?

☐ excellent   ☐ very good   ☐ good

(a) Write down one thing that is wrong with this question. **(1)**
(b) Write an improved question. You must include response boxes. **(2)**

You can comment on either the question or the response boxes.

Capture/recapture

You can use the capture/recapture method to estimate population size for large populations. The diagrams below show how this method works on a population of fish.

1 Catch a sample of fish and mark them. Record how many have been marked and return them.

The experimenter catches 4 fish, marks them and returns them.

LATER

2 Catch a second sample of fish and count how many in this sample are marked.

The experimenter catches 5 fish in a second sample. 2 are found to be marked.

You can compare **ratios** to estimate the size of the population. You **assume** that these two ratios are equivalent:

Ratio of marked fish : total fish in recapture sample

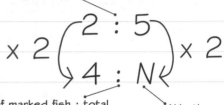

Ratio of marked fish : total fish in whole population

N is the population size

Using a formula

You can use this formula to estimate the population size, N:

$$N = \frac{Mn}{m}$$

M = number of fish marked then released
n = size of recapture sample
m = number of marked fish in recapture sample

Worked example

tier **H**   Aiming higher

Another method is to compare the ratios:
$6 : 40 = 60 : N$

A scientist wants to estimate the number of mice on an island. She captures 60 mice, marks them and releases them. Later she captures 40 mice, of which 6 were already marked.

(a) Find an estimate for the number of mice on the island. **(2)**

If the number of mice on the island is N, then

$$\frac{N}{60} = \frac{40}{6} \text{ so } N = \frac{60 \times 40}{6} = 400$$

(b) Write down two assumptions the scientist has to make. **(2)**

1. The marked mice do not die.
2. The probability of being caught is the same for marked and unmarked mice.

You should give any assumptions in context. Other possible answers are that the mice cannot escape off the island and that there are no births or deaths.

Assumptions

Learn the underlying assumptions.

✓ The population is closed – no migration.

✓ All members of the population are equally likely to be capture in each sample.

✓ Capture and marking do not affect catchability.

✓ The population does not change due to deaths or births between sampling occasions.

Now try this

tier **H**

1 In a study of polar bears in the Barents Sea, 39 bears were captured, marked and released. One year later a sample of 120 bears yielded 9 that were marked.

(a) Work out an estimate for the number of polar bears in the Barents Sea. **(2)**
(b) Write down one assumption that has to be made. **(1)**

Frequency tables

Frequency tables can be used to summarise both quantitative and qualitative information.

Frequency tables for qualitative data

Here is a frequency table which has been used to summarise qualitative information.

The table gives information about the types of wood that chairs in a shop are made of.

There were four chairs made of ash wood.

Type of wood	Tally	Frequency
Ash	////	4
Oak	//// //	7
Pine	////	5
Beech	////	4
		20

For the 5th tally mark, put a line through the previous 4.

The total number of chairs goes here.

Grouped frequency tables for quantitative data

When grouping quantitative data, the class intervals must not have gaps in between them or overlap each other.

The list shows the ages (in years) of 15 people.

8 6 5 10 15 18 28 32
38 23 11 38 26 27 21

These are the class intervals.

Age, A (years)	Tally	Frequency
$0 < A \leq 10$	////	4
$10 < A \leq 20$	///	3
$20 < A \leq 30$	////	5
$30 < A \leq 40$	///	3

Be careful with $<$ and $\leq$.
The value of 10 years is tallied in the first row, not the second.

Making a grouped frequency table

1 Decide on the class intervals.

2 Make a three-column table with 'Tally' and 'Frequency' in the 2nd and 3rd columns.

3 Write the heading for the 1st column and then the class intervals, taking care over the $<$ and $\leq$ signs.

4 Cross off the ages in the list as you tally.

5 Fill in the frequency column and then check that the total is the same as the number of ages in the list.

Worked example

tier F&H

The table gives some information about the number of TVs in each house on a street.

Number of TVs in house	Frequency
0	1
1	8
2	9
3	7

(a) Work out the total number of houses. **(1)**

$1 + 8 + 9 + 7 = 25$

(b) Work out the total number of TVs. **(2)**

$0 \times 1 + 1 \times 8 + 2 \times 9 + 3 \times 7 = 47$

 Watch out! 0×1 is 0 not 1.

EXAM ALERT!

Lots of students make the mistake of just adding up the numbers in both columns.

For part (a) add up the numbers in the frequency column.

For part (b) multiply the number of TVs in each house by its frequency, then add them up.

Students have struggled with exam questions similar to this – **be prepared!** Results Plus

Now try this

1 Here are the lengths of time (in seconds) that some people could hold their breath:
8 12 25 18 30 23 35 33 28 18 9 18 22 10 21 15 29 31 20 23
Record this information in a grouped frequency table with intervals of equal width. **(3)**

Two-way tables

A two-way table shows information about two categories of data.

Some values in a two-way table will be given; others may need to be calculated using addition or subtraction.

This two-way table shows people's replies when they were asked if they preferred sponge pudding with custard or with cream.

No one is allowed to pick both custard and cream in their reply.

34 adults preferred custard.

	With custard	With cream	Total
Adult	34	18	52
Child	22	23	45
Total	56	41	97

The figures in blue show the replies.
The figures in red are worked out from the table.

Completing a two-way table

This two-way table shows the numbers of male and female musicians in each section of an orchestra.

To complete a two-way table look for rows or columns with only one missing value.

Top row total is 23 + 17 = 40

	Strings	Wind and brass	Total
Male	23	17	40
Female	34	8	42
Total	57	25	82

42 − 34 = 8 so '8' must go in female wind and brass.

Worked example

tier F

A group of 53 students was asked what their main source of news was – TV or the internet.
There were 32 boys and 21 girls in the group.
31 students, of which 18 were boys, said TV.
(a) Complete the two-way table. **(2)**

	TV	Internet	Total
Boys	18	14	32
Girls	13	8	21
Total	31	22	53

(b) What was the students' main source of news? **(1)**
TV. The table shows that 9 more students used TV for their main source of news than used the internet.

In this case you have to fill in a two-way table from the information given. Make sure you use all the information given in the question.

Begin by putting in the values you are given. Don't forget the 53 in the first sentence.

Look for rows or columns with a single gap, and work out the missing numbers.

Check that the numbers in the rows and in the columns add up to the totals.

Always give your answer in context and with a reason.

Now try this

tier F

1 A teacher asked 30 students if they had a school lunch, a packed lunch or if they went home for lunch.
17 of the students were boys. 4 of the boys had a packed lunch. 7 girls had a school lunch.
3 of the 5 students who went home were boys.

(a) Record this information in a two-way table. Use the headings 'School', 'Packed' and 'Home'. **(3)**
(b) Work out the number of students who had a packed lunch. **(1)**

Pictograms

Pictograms are a way of summarising data in a chart. You need to be able to draw and interpret pictograms.

Pictograms

This pictogram shows information about sales from a shop.

You will be expected to deal with halves and quarters when interpreting pictograms.

Pictograms may not be suitable for large numbers as symbols often cannot be easily divided into smaller than quarters.

Week 1	🍾 🍾 🍾	← 3 × 12 = 36 bottles sold in week 1.
Week 2	🍾 🍾	
Week 3	🍾 🍾 🍾 🍾	← This half symbol means 12 ÷ 2 = 6 bottles. So 42 bottles were sold in week 3.
Week 4	🍾	

Key 🍾 represents 12 bottles ← The **key** shows how many each symbol stands for.

Worked example

tier F

The pictogram gives some information about the numbers of cheeses Pippa sold in her shop in March, in April and in May.

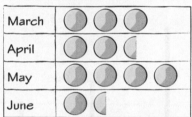

March	◐ ◐ ◐
April	◐ ◐ ◐
May	◐ ◐ ◐ ◐
June	

Key ◐ represents 10 cheeses

Pippa sold 30 cheeses in March.
(a) Complete the key. **(1)**
Pippa sold 15 cheeses in June.
(b) Complete the pictogram. **(1)**
The price of a cheese is £18.
(c) Work out the total price of the cheeses Pippa sold in May and June. **(1)**

4 circles in May gives 40 cheeses.
40 + 15 = 55
Total price = 55 × £18 = £990

> Use the information given to work out the key. Pippa sold 30 cheeses in March and there are 3 shapes in the 'March' row, so 1 shape represents 10 cheeses.

> A full shape represents 10 cheeses so a half shape represents 5 cheeses. Draw 1 full shape and 1 half shape to show 15 cheeses.

> You need to be prepared to use the information given in a pictogram to carry out further calculations. Add together the numbers of cheeses sold in May and June then multiply by £18.

Now try this

tier F

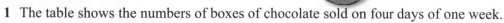

1 The table shows the numbers of boxes of chocolate sold on four days of one week.

Monday	Tuesday	Wednesday	Thursday
4	12	6	10

Draw a pictogram for this information. Use the symbol ▦ to represent four boxes. **(3)**

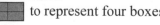

Bar charts and vertical line graphs

Bar charts and vertical line graphs are a good way of representing **discrete** data given in a tally chart or frequency table.

They can also be used to represent qualitative data. You met these types of data on page 1

Bar charts

The table shows information about the types of trees in a wood.

Type of tree	Frequency
Oak	6
Ash	8
Pine	4
Beech	2

The bar chart shows how this information can be displayed.

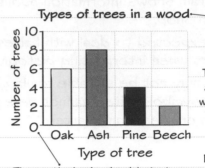

Types of trees in a wood — The bar chart should have a title.

The bars should all be the same width and equally spaced.

The vertical axis should start from 0 and go up by equal amounts each division.

Both axes must be labelled. You can label the vertical axis 'Frequency'.

Worked example

tier **F**

This vertical line graph (stick graph) gives information on the town people prefer to shop in.

(a) How many people said Bath?   **(1)**

12

(b) More people preferred Bristol than Swindon. How many more?   **(1)**

$15 - 8 = 7$

(c) How many people were asked altogether?   **(1)**

$12 + 15 + 10 + 8 = 45$

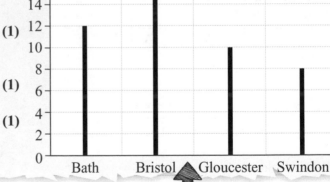

Which town people preferred to shop in

The line that goes up to between 14 and 16 must be a whole number – so 15.

This type of graph is similar to a bar chart but it has very thin bars.

Now try this

tier **F**

1 The table gives information about the numbers of vehicles on a road one morning.

Car	Van	Lorry	Bus	Other
8	7	11	3	5

Draw a bar chart to show this data.

Make sure you start the scale at 0 and go up in equal steps.   **(2)**

Stem and leaf diagrams

Stem and leaf diagrams are commonly used to display small amounts of discrete individual data.

Drawing a stem and leaf diagram

A stem and leaf diagram shows data in an ordered way.

The diagram shows the numbers of emails 15 people received one day.

The **key** is necessary to interpret the diagram. The **leaves** must always be single figures.

The column with 0, 1, 2 and 3 is the stem.

The rows contain the leaves.

In this case the tens make the stem and the units are the leaves.

0	6	8			
1	2	2	4	6	
2	0	2	3	4	7	7
3	1	3	5		

Key 3|5 means 35 emails

Worked example

tier F

Here are the numbers of goals a player has scored in her last 20 netball games.

23 34 20 14 23 6 17 24 24 18
16 10 22 21 33 8 21 15 8 22

Draw an ordered stem and leaf diagram to show this data. **(3)**

0	6	8	8					
1	4	7	8	6	0	5		
2	3	0	3	4	4	2	1	1	2
3	4	3						

0	6	8	8					
1	0	4	5	6	7	8		
2	0	1	1	2	2	3	3	4	4
3	3	4						

Key 3|4 means 34 goals

Then copy the stem and rewrite the leaves on each row in order of increasing size.

Top tips

When you draw a stem and leaf diagram, **always**:

1 Look for the smallest and the largest numbers.

2 Draw the vertical stem from smallest to largest.

3 Write down the leaves (rows) in order from the list of data.

4 Then redraw the diagram with the leaves in order.

Draw a vertical line and put the numbers 0 to 3 in a column on the left of it. Next draw the horizontal rows and fill them in with the units. Cross off each number in the original list as you copy it onto the diagram.

EXAM ALERT!

Students often lose marks by failing to write a correct key. Make sure you explain what values your stem and your leaves represent. Remember to include any units such as cm, kg, goals, people.

Students have struggled with exam questions similar to this – **be prepared!**

Results Plus

Now try this

tier F&H

1 Here are the heights, in cm, of 16 stalks of wheat. Show the heights on a stem and leaf diagram.
120 131 108 110 122 132 127 105
122 133 121 137 112 119 104 113   **(2)**

2 Here are the lengths, in cm, of 15 worms.
3.8 4.7 5.6 7.2 6.4 3.9 4.7 4.1
5.3 6.1 7.1 6.3 5.5 4.9 6.0
Show the lengths in a stem and leaf diagram. **(2)**

Start the stem with 10, 11 and so on. The leaves should always be single figures.

Pie charts

Pie charts are generally used to show **qualitative** data. You need to be able to interpret them accurately. Remember that the angle of any sector in a pie chart is **proportional** to the number it represents.

Interpreting pie charts

This pie chart gives information about the replies that students gave to the question 'What is your most important subject?'

The pie chart shows that 'English' got the most votes and 'Maths' got one quarter of the votes.

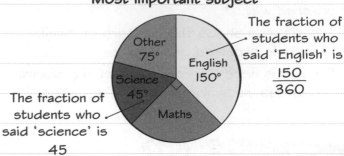

Most important subject

The fraction of students who said 'English' is $\frac{150}{360}$

The fraction of students who said 'science' is $\frac{45}{360}$

Using a formula

You can use this rule to work out what each sector represents:

$$\text{number represented} = \frac{\text{angle of sector}}{360} \times \text{total}$$

Worked example

tier F&H

The pie chart gives information from a survey about where people worked.
There were 135 people in the survey.

Place of employment

Other 40°
Swindon
Bath 128°
West Wilts 96°
Bristol 64°

(a) Work out how many people worked in Bath. **(2)**

For Bath $\frac{128}{360} \times 135 = 48$ people

(b) Work out how many people worked in Swindon. **(2)**

Angle for Swindon $= 360° - 128° - 64° - 96° - 40° = 32°$

Number for Swindon $= \frac{32}{360} \times 135 = 12$

Golden rule

Use $\frac{\text{angle}}{360} \times \text{total}$ to find the number given by each sector.

EXAM ALERT!

You should also be able to measure angles on pie charts and use your measured angles in the formula above.

Students have struggled with exam questions similar to this – **be prepared!**

Remember that the sum of the angles in a pie chart must be 360°.

Now try this

tier F&H

1 The pie chart gives information about the masses of vegetables grown on a farm.
   The total mass of vegetables produced was 2000 kg.
   Work out the mass of potatoes produced. **(2)**

Production of vegetables

Other 36°
Leeks 72°
Potatoes 72°
Marrows 144°
Beans 36°

Drawing pie charts

You need to be able to draw pie charts accurately from given data. You will need a pair of compasses, a protractor and a ruler.

Working out angles for pie charts

The sizes of the sectors in a pie chart are proportional to the numbers they represent.

Blood type	O	A	B	AB
Number	30	24	12	6

Use this rule to work out the angles for the pie chart.

$$\text{angle} = \frac{\text{number to be represented}}{\text{total}} \times 360°$$

Total = 30 + 24 + 12 + 6 = 72

Angle for blood type O = $\frac{30}{72} \times 360 = 150°$

Angle for blood type A = $\frac{24}{72} \times 360 = 120°$

Angle for blood type B = $\frac{12}{72} \times 360 = 60°$

Angle for blood type AB = $\frac{6}{72} \times 360 = 30°$

Check
150° + 120° + 60° + 30° = 360° ✓

Worked example

tier **F**

The table shows the types of flats that were for sale in a town last year.

Type	Bedsit	1-bed	2-bed	3-bed
Number	90	42	30	18

Draw a pie chart to show this information. **(3)**

Total = 90 + 42 + 30 + 18 = 180 flats

So the angle for bedsits = $\frac{90}{180} \times 360 = 180°$

The other angles are 84°, 60°, 36°

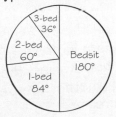

Types of flat for sale

Work out the total first.
The angle for 'Bedsits' = $\frac{90}{\text{total}} \times 360°$

Work out the other angles.

Use a pair of compasses, a protractor and a ruler to draw the sectors.

Don't forget to **label** the sectors.

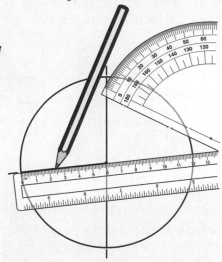

Now try this

tier **F**

1 The table shows the types of dairy cow on a large farm.

Type of cow	Holstein	Friesian	Jersey	Ayrshire
Number	80	40	36	24

Draw a pie chart to show this information. **(3)**

Bar charts

Multiple bar charts have more than one bar for each class.

Composite (or compound) bar charts group several different bars into a single bar and often show percentages.

Multiple bar charts

This multiple bar chart shows the sales of three makes of cars in four quarters of one year.

There are four sets of three bars to show how sales change over the year.

Using this bar chart it is easy to see that sales of Seat cars were high in the first two quarters but then fell.

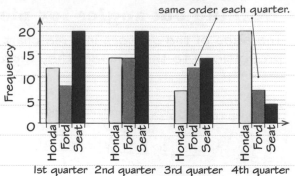

The bars must be in the same order each quarter.

Composite bar charts

This composite bar chart shows how the percentages of men and women seen jogging have changed over two years.

The percentage of women has increased from 25% to 40%.

Composite bar charts can be harder to understand than multiple bar charts but do show the proportions within each group better.

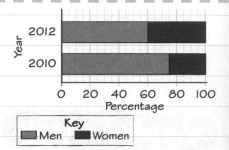

Key
Men   Women

Worked example

tier F&H

The composite bar chart shows the percentages of families with 1, 2 and 3 children in a town. Did the percentage of families with 2 children increase or decrease between 1980 and 2010? **(2)**

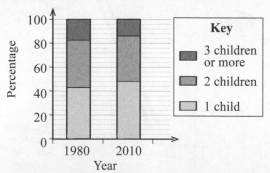

Key

3 children or more

2 children

1 child

In 1980, the percentage was 82 − 43 = 39
In 2010, the percentage was 86 − 48 = 38
So the percentage decreased.

You need to be careful when you are reading off a composite bar chart. Use the key to work out which rectangle represents two children. If this rectangle doesn't start at 0 you will have to read off the values at the top and bottom of the rectangle and use subtraction.

Read off the percentage of families with 2 children in 1980 and in 2010. Write down both values, then write a conclusion saying whether the percentage increased or decreased.

Now try this

tier F

1 The table gives the times (in minutes) two boys spent watching TV on four days of one week.

Monday		Tuesday		Wednesday		Thursday	
Keith	Andy	Keith	Andy	Keith	Andy	Keith	Andy
40	30	30	50	85	70	140	170

(a) Draw a multiple bar chart to show this information. **(3)**
(b) Describe the trend in the times Andy spent watching TV. **(1)**

Don't forget to include a key.

Pie charts with percentages

Pie chart sectors can be labelled with percentages when comparing data sets of different sizes.

Percentages and pie charts

In a pie chart the sectors could be labelled with percentages rather than angles or numbers.

BBC Radio channel

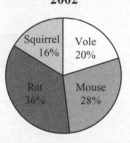

The angle for Radio 2 is 25 × 3.6 = 90°

Golden rules

1 Percentages can be turned into angles using

$$\text{angle} = \frac{\text{percentage}}{100} \times 360$$
$$= \text{percentage} \times 3.6$$

2 Angles can be turned into percentages using

$$\text{percentage} = \text{angle} \div 3.6$$

3 You can find the number represented by a sector using

$$\text{number} = \text{total} \times \text{percentage} \div 100$$

Worked example

tier F&H

The pie charts show data about four types of rodents found in two different years.

2002   **2012**

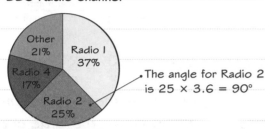

The total number of rodents in 2002 was 500.

(a) Work out the number of rats in 2002. **(2)**

$$500 \times \frac{36}{100} = 180$$

(b) Calculate the angle that represents squirrels in the 2012 pie chart. **(1)**

Angle = 15 × 3.6 = 54°

Clare says that the number of rats has fallen between 2002 and 2012. Clare could be wrong.

(c) Explain why. **(1)**

The pie charts only show percentages. The total number in 2012 is unknown.

EXAM ALERT!

Students often answer questions like part (c) poorly. Remember that pie charts describe **proportions**, not total numbers. Although the **percentage** of rats has **decreased**, their **total number** could have **increased** if the overall number of rodents had increased substantially.

Students have struggled with exam questions similar to this – **be prepared!**   Results**Plus**

Use the third golden rule.

Use the first golden rule.

The pie charts show percentages, so a comparison of proportions can be easily made for the two years. However, the pie charts do not show total numbers.

Now try this

tier F&H

Remember to use the first golden rule.

1 The table shows the percentages of flower bulbs sold from a garden centre.

Type of bulb	Crocus	Daffodil	Tulip	Hyacinth
Percentage (%)	25	35	30	10

Draw a pie chart to show this information. **(3)**

Using comparative pie charts

When you have to compare the numbers in populations of different sizes, you can use pie charts of different sizes.

Comparative pie charts

Here are two pie charts showing men's and women's preferred sources for news.

The pie charts are different sizes because the total number of men in the sample is different from the total number of women.

Men

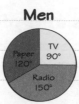

Women

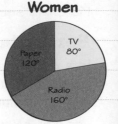

$$\frac{\text{area of large pie chart}}{\text{area of small pie chart}} = \frac{\text{total number of women}}{\text{total number of men}}$$

Making a comparison

Total number = n Total number = N

$$\frac{\text{area of large pie chart}}{\text{area of small pie chart}} = \frac{\pi R^2}{\pi r^2} = \frac{R^2}{r^2} = \frac{N}{n}$$

So, use $\frac{R^2}{r^2} = \frac{N}{n}$ where r is the radius of the small circle, R is the radius of the large circle, n is the total number in the small sample and N is the total number in the large sample.

Worked example

Aiming higher · **tier H**

Holiday destinations 2012

Radius = 3 cm

Holiday destinations 2013

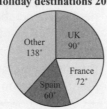

Radius = 4.5 cm

The pie charts show information about the holiday destinations of a sample of 200 people in 2012 and a sample of people in 2013.

Work out how many more people went to France in 2013 than in 2012. **(3)**

The total number of people in 2013 = N

$\frac{R^2}{r^2} = \frac{N}{n}$ so $\frac{4.5^2}{3^2} = \frac{N}{200}$

$N = \frac{20.25}{9} \times 200 = 450$

Number who went to France in 2012

$= \frac{72}{360} \times 200 = 40$

Number who went to France in 2013

$= \frac{72}{360} \times 450 = 90$

So 50 more people went to France in 2013 than in 2012.

You only need to change the radius if the pie charts are going to be used for comparison.

You will need to remember the formula $\frac{R^2}{r^2} = \frac{N}{n}$ which relates the two radii and the two sample sizes.

Remember: number = $\frac{\text{total} \times \text{angle}}{360}$

Although the angle for France was the same in both pie charts, the number it represents is not because the pie charts are **different sizes**.

Now try this

tier H · **Aiming higher**

1 John is studying the distribution of weeds in marshland and in moorland.
   The size of the marshland sample is 200 and the size of the moorland sample is 450.

   John draws a pie chart of radius 4 cm for the marshland.
   What radius should he use for the moorland? **(1)**

Frequency polygons

A frequency table is a table for summarising information.
You can use a frequency table to draw a frequency polygon to show the information.

Frequency tables

You can use frequency tables for grouped **discrete** data or for grouped **continuous** data.
A length of 10 cm **would not** go in the interval $10 < L \leqslant 20$, but a length of 20 cm **would**.

There should be a gap between successive class intervals so there is no overlap.

Discrete data

Number of cars	0–5	6–10	11–15	16–20
Frequency	5	7	8	10

Use $20 < L \leqslant 30$ and $30 < L \leqslant 40$ to ensure no gaps and no overlaps.

Continuous data

Length, L (cm)	$0 < L \leqslant 10$	$10 < L \leqslant 20$	$20 < L \leqslant 30$	$30 < L \leqslant 40$
Frequency				

A length of exactly 20 cm would go in this class interval.

Frequency polygons

You can use a frequency polygon to show data from frequency tables.

Frequencies are plotted in the **middle** of each interval.

Lifetime, H (hours)	$0 < H \leqslant 1$	$1 < H \leqslant 2$	$2 < H \leqslant 3$	$3 < H \leqslant 4$
Frequency	7	10	5	2

The points are plotted in the middle of each interval.

The first and last points are not joined elsewhere.

Points are joined by straight lines.

Find the midpoints by adding the two end values and dividing by 2.

Worked example

tier F&H

The table shows information about the lengths of some flower stalks.

Length, L (cm)	$0 < L \leqslant 10$	$10 < L \leqslant 20$	$20 < L \leqslant 30$	$30 < L \leqslant 40$
Frequency	7	12	18	7

Draw a frequency polygon to show this information. (2)

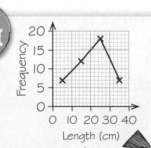

The midpoints are at 5, 15, 25, 35.
Plot the points at (5, 7), (15, 12), (25, 18), (35, 7).
Join the points with straight lines using a ruler.

Check: The crosses should be equally spaced horizontally.

Now try this

tier F&H

1 The table gives information about the times of flight, t minutes, of some hot air balloons.

Time, t (mins)	$0 < t \leqslant 10$	$10 < t \leqslant 20$	$20 < t \leqslant 30$	$30 < t \leqslant 40$	$40 < t \leqslant 50$
Frequency	6	9	8	7	5

The last midpoint will be at 45.

Draw a frequency polygon to show this information. (2)

23

Cumulative frequency diagrams 1

You have to know how to produce a cumulative frequency diagram from a frequency table.

Cumulative frequency

Here is a frequency table with continuous data showing the times (in minutes) people waited for a phone call to be answered.

Time, t (min)	$0 < t \leq 1$	$1 < t \leq 2$	$2 < t \leq 3$	$3 < t \leq 4$
Frequency	5	10	20	5

Here is the **cumulative** frequency table for the same data.

Time, t (min)	$0 < t \leq 1$	$0 < t \leq 2$	$0 < t \leq 3$	$0 < t \leq 4$
Cumulative Frequency	5	5 + 10 = 15	15 + 20 = 35	35 + 5 = 40

35 people had to wait for up to 3 minutes.

The total number of people was 40.

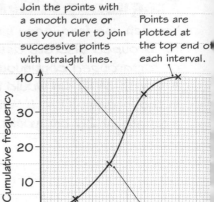

Join the points with a smooth curve **or** use your ruler to join successive points with straight lines.

Points are plotted at the top end of each interval.

This is (1,5) from the first interval.

Ensure the curve through all the po

The table gives information about the heights of 100 students.

Height, h (cm)	Frequency	Cumulative frequency
$120 < h \leq 130$	8	8
$130 < h \leq 140$	16	24
$140 < h \leq 150$	24	48
$150 < h \leq 160$	32	80
$160 < h \leq 170$	20	100

Draw a cumulative frequency diagram to show this information. **(3)**

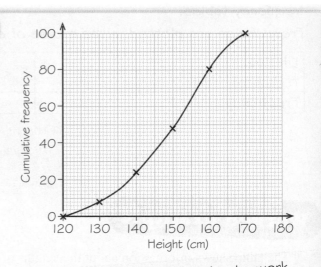

EXAM ALERT!

Many students make the mistake of plotting points in the middle of the intervals rather than at the top end.

Students have struggled with exam questions similar to this – **be prepared!**

ResultsPlus

Add together the frequencies to work out the cumulative frequencies. Then plot the cumulative frequency at the **top end** of each class interval. Be careful with the scale on the vertical axis. Each large grid square represents 10 students, so each small grid square represents 2 students.

Now try this

tier **F&H**

1 The table shows some information about the snowfall each day in January and February one year.

Snowfall, S (cm)	$0 < S \leq 2$	$2 < S \leq 4$	$4 < S \leq 6$	$6 < S \leq 8$	$8 < S \leq 10$
Frequency	20	14	12	8	6

Draw a cumulative frequency diagram to show this information. **(3)**

Start by workin out the cumulat frequencies.

Cumulative frequency diagrams 2

You need to be able to interpret cumulative frequency diagrams.

Drawing cumulative frequency diagrams is described on page 24. Using a cumulative frequency diagram to find the median is described on page 37.

Cumulative frequency step polygons

You can use a cumulative frequency step polygon to represent **discrete** data given in a frequency table.

Discrete data can only take certain values, often whole numbers.

The table shows the numbers of cars in a car park at noon on 38 days.

Number of cars	11	12	13	14	15
Frequency	8	11	9	6	4

The cumulative frequency remains at 8 until 12 cars is reached. It then jumps up to 19.

The cumulative frequency is zero until 11 cars is reached. It then jumps to 8.

Worked example
tier F&H

The diagram gives some information about the costs of some laptops.

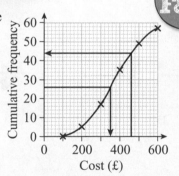

Draw a vertical line from 460 on the horizontal axis to the curve.
Read off the value on the cumulative frequency axis (44).
Remember to subtract from the total number (57).

(a) Find an estimate for the number of laptops with a cost of more than £460. **(1)**

57 − 44 = 13

(b) Find an estimate for the cost which the cheapest 26 laptops do not exceed. **(1)**

£350

Draw a horizontal line from 26 across to the graph. Read off the value from the cost axis.

Now try this
tier F&H

1

The diagram gives the cumulative frequency percentages of ages of people in India.

(a) Find an estimate for the percentage of people who are aged 20 or less. **(1)**

(b) What age is exceeded by 25% of the people in India? **(1)**

Remember to use a ruler and sharp pencil when you draw lines for reading off.

Histograms with equal intervals

You use a histogram to represent **grouped continuous data**. There are **no gaps** between the bars on a histogram (unless one of the intervals has a zero frequency).

Interpreting histograms

You can read frequencies from the histogram as well as work out cumulative frequencies.

This histogram shows the distances people could throw a rock.

It shows that there were 5 people who threw between 4 m and 5 m.

You can work out that there were 5 + 8 = 13 people who threw up to 6 m.

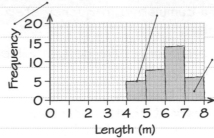

Frequency is always on the vertical axis.

5 people threw between 4m and 5m.

In histograms with equal intervals, the bars **must** be equal width.

Worked example

tier F&H

The table gives some information about the heights of some children in a club.

Height, h (cm)	Frequency
$110 < h \leqslant 120$	8
$120 < h \leqslant 130$	13
$130 < h \leqslant 140$	16
$140 < h \leqslant 150$	10
$150 < h \leqslant 160$	7

Draw a histogram to show this information. **(2)**

Draw each bar to the given frequency. Use a ruler and a sharp pencil whenever you are drawing graphs in the exam.

Now try this

tier F&H

1

Width, W (cm)	$0 < W \leqslant 2$	$2 < W \leqslant 4$	$4 < W \leqslant 6$	$6 < W \leqslant 8$	$8 < W \leqslant 10$
Frequency	12	9	7	4	2

Draw a histogram to show the information in the table. **(2)**

2

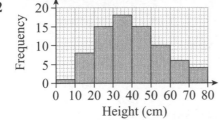

Remember that there must be **no** gaps between the bars.

The histogram gives information about the heights of plants in a garden.

How many plants had a height of 40 cm or greater? **(2)**

Histograms with unequal intervals

Histograms with **unequal** class intervals use **frequency density**, not frequency.

The graph shows the times taken by some students to solve a problem. The number of students that a bar represents is equal to the **area** of the bar.

The first bar shows that 8 (1 × 8) students took less than 1 minute. The fourth bar shows that 7 (0.5 × 14) students took between 3.5 and 4 minutes.

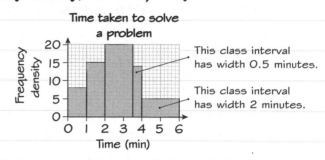

This class interval has width 0.5 minutes.

This class interval has width 2 minutes.

Frequency density

Before drawing a histogram you need to calculate the **frequency density** for each class interval.

The table shows the lengths of leaves in a garden.

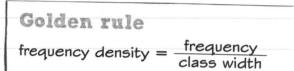

Golden rule

$$\text{frequency density} = \frac{\text{frequency}}{\text{class width}}$$

Length, L (cm)	Frequency	Class width	Frequency density
$0 < L \leqslant 5$	12	5	12 ÷ 5 = 2.4
$5 < L \leqslant 15$	10	10	10 ÷ 10 = 1.0
$15 < L \leqslant 25$	16	10	16 ÷ 10 = 1.6
$25 < L \leqslant 45$	12	20	12 ÷ 20 = 0.6
$45 < L \leqslant 60$	6	15	6 ÷ 15 = 0.4

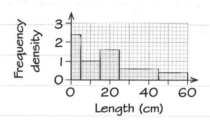

Class interval = 45 − 25 = 20

Worked example

tier H

The table gives data about the times some trains were late.

Time, t (min)	Frequency	Class width	Frequency density
$0 < t \leqslant 5$	12	5	12 ÷ 5 = 2.4
$5 < t \leqslant 10$	18	5	18 ÷ 5 = 3.6
$10 < t \leqslant 20$	14	10	14 ÷ 10 = 1.4
$20 < l \leqslant 40$	12	20	12 ÷ 20 = 0.6

Draw a histogram to show this data.   **(3)**

1. Work out each class width.
2. Use the golden rule to work out each frequency density.

3. Label the vertical axis Frequency density.
4. Decide on the scale for each axis (these may be given to you in the exam).
5. Draw the bars.

Now try this

tier H

1 The table gives information about the calorie intake per day of some people.

Number (N) of calories	$1000 < N \leqslant 1750$	$1750 < N \leqslant 2000$	$2000 < N \leqslant 2250$	$2250 < N \leqslant 2750$	$2750 < N \leqslant 4000$
Frequency	16	34	42	54	14

Draw a histogram to show this information.

UXBRIDGE COLLEGE
LEARNING CENTRE

(3)

Interpreting histograms

You can find frequencies from histograms either by using frequency density or by using an area key.

Histograms and area keys

Histograms can be drawn or interpreted by using an **area key**.

In this histogram, one small square represents $50 \div 25 = 2$ fish.

The number represented by each bar in the histogram can be found by using the area key.

For example, the number of fish with length 15 cm or less is given by the area of the first two bars.

Area $= 10 \times 5 + 16 \times 2.5 = 90$ small squares

Number of fish $= 90 \times 2 = 180$

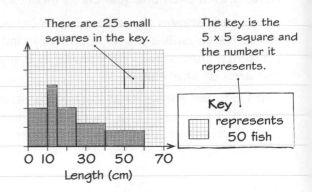

There are 25 small squares in the key.

The key is the 5 x 5 square and the number it represents.

Key
☐ represents 50 fish

Length (cm)

Worked example

tier **H**   Aiming higher

The histogram shows the times (in seconds) people spent at a self-service checkout.

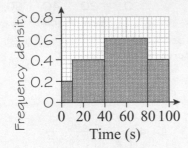

Time (s)

Time, t (s)	Frequency
$0 < t \leqslant 10$	2
$10 < t \leqslant 40$	12
$40 < t \leqslant 80$	24
$80 < t \leqslant 100$	8

Use the histogram to complete the frequency table.

The frequency density of the $80 < t \leqslant 100$ class interval is $\frac{8}{20} = 0.4$

Frequency of the $0 < t \leqslant 10$ class $= 0.2 \times 10 = 2$

Frequency of the $10 < t \leqslant 40$ class $= 0.4 \times 30 = 12$

Frequency of the $40 < t \leqslant 80$ class $= 0.6 \times 40 = 24$

Another method is to use the information you have been given to create an area key

 represents 4 people

Use the frequency given to work out the vertical scale.

Frequency = frequency density × class width

Now try this

tier **H**   Aiming higher

1 The histogram gives information about the water intake per day of some plants.

Water intake, W (ml)	$100 < W \leqslant 150$	$150 < W \leqslant 200$	$200 < W \leqslant 300$	$300 < W \leqslant 500$
Frequency	200	300	450	540

(a) Draw a histogram to show this infomation. Use a horizontal scale of 1 cm = 100 ml. Use an area key of 1 cm² represents 200 plants. **(4)**

(b) Estimate the percentage of plants which had an intake of more than 250 ml per day. **(3)**

Population pyramids

A population pyramid consists of a pair of back-to-back histograms.

It gives information about the age structure of a population.

The histograms are presented horizontally with male and female populations on opposite sides.

Population pyramids

The diagram shows the population (in thousands) of males and females grouped by age.

In the 21–40 age group there are 135 000 males and 125 000 females.

This diagram shows that the majority of the population are young (because the longest bars are at the bottom).

The vertical scale always has the youngest ages at the bottom.

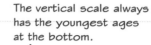
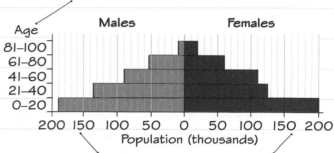

The scales on sides must be the same.

Worked example

tier F&H

Read off the percentages for males aged 0–20 and males aged 21–40 then add them together.

The population pyramid shows the distribution of males and females in a population as a percentage of the total number of each gender.

(a) What percentage of males are 40 or under? **(1)**

35 + 25 = 60%

(b) What percentage of females are over 40? **(1)**

17 + 14 + 11 = 42%

Read off the percentages for females aged 41–60, then 61–80 and then 81–100 and add them together.
Write down each percentage as you read it off.

The values shown by the horizontal bars on each side should sum to 100%, allowing for a small error caused by the rounding of the numbers for each bar.

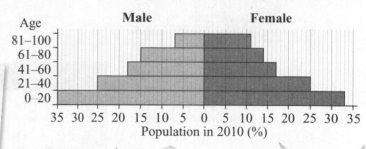

Now try this

tier F&H

Make two different comments comparing the percentages of the population in different age groups. Give actual percentages to back up your comments. You can use **skew** to describe the distributions – there is more about this on page 48.

1
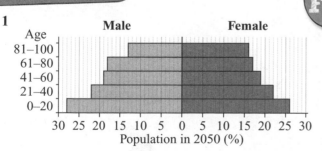

Look at the population pyramid in the worked example above and this population pyramid, which shows the predicted distribution of the population in 2050.

Compare the population in 2010 with the predicted population in 2050. **(2)**

Choropleth maps

The distribution of the surname Davies in the UK

Choropleth maps

A choropleth map uses different colours or shading to show how data varies across different **geographic** areas. This choropleth map shows the distribution in the UK of the surname 'Davies'.

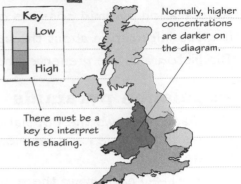

Key

Low

High

Normally, higher concentrations are darker on the diagram.

There must be a key to interpret the shading.

Choropleth maps on grids

You can use a choropleth map to display the change in population over an area.

This diagram shows the masses of worms that live in the topsoil of an 8 m by 12 m plot.

In your exam, choropleth maps will use different shades of grey or shading patterns to show different values.

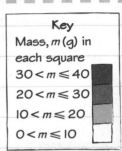

Key

Mass, m (g) in each square

$30 < m \leqslant 40$

$20 < m \leqslant 30$

$10 < m \leqslant 20$

$0 < m \leqslant 10$

Worked example

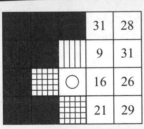

	31	28
	9	31
○	16	26
	21	29

Key: Number of flowers

31–40
21–30
11–20
0–10

The diagram shows a field split into 2-metre squares. The numbers in each square show the numbers of flowers.

(a) On the grid provided complete the choropleth. **(2)**

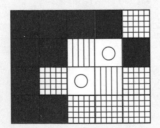

You will be given a key showing how to shade for different values.

You can make any general comment which describes the spread.

tier **F**

(b) Describe the distribution of the flowers. **(1)**
There is a higher density of flowers on the edges of the field.

EXAM ALERT!

Some students lose marks because they just wri[te] down some of the numbers given in the diagram instead of a general comment.

Students have struggled with exam questions similar to this – **be prepared!**

ResultsPl[us]

Now try this

tier **F**

1

140	139	311	312
126	320	211	189
330	326	201	116
300	270	170	80

See what the highest and lowest values are before deciding on a key. Then think of suitable intervals – it's usual to have four.

Remember to give a general comment.

The diagram shows a square field. The numbers in the small squares show the numbers of insects.
(a) Draw a suitable choropleth map. You must include a suitable key. **(2)**
(b) Describe the distribution of the insects. **(1)**

Mode, median and mean

The **mode**, **median** and **mean** are the three averages you need to know for your exam.

The **mode** is the value appearing **most** often.

The **median** is the **middle** value when they are in order.

To work out the **mean**, add up all the values and divide by how many values there are.

The mode is the height of any of these people.   The median is the height of this person in the middle.

In order of increasing height

Worked example

tier F

Here is a list of the numbers of children in nine families.

6 3 2 2 3 1 1 8 1

(a) Find the mode.   **(1)**

In order: 1 1 1 2 2 3 3 6 8

The mode is the most common so the mode is 1.

(b) Find the median.   **(1)**

The median is the middle value when written in order so the median is 2.

Order the data from smallest to largest. Make sure you have the same number of values in the ordered list as in the original list.

You can use the rule $\frac{n+1}{2}$ to find where the middle value of n values is. There are 9 data values. $\frac{9+1}{2} = 5$, so the 5th value is the mode.

Median

You need to be careful when you are calculating the median of an **even** number of data values.

You can still use the rule that the median is the $\frac{n+1}{2}$th value.

Or you can cross off the first and the last, the second and the second last and so on until you get to the middle two values, then calculate the value halfway between them.

Here is a list of the number of cars in a small car park over a period of 14 days.

14 16 18 15 14 18 13 20 22 19 12 12 21 25

In order:

12 12 13 14 14 15 16 18 18 19 20 21 22 25

$\frac{n+1}{2} = \frac{14+1}{2} = 7.5$

The 7.5th value is halfway between the 7th and 8th values.

1̷2̷ 12 1̷3̷ 1̷4̷ 1̷4̷ 1̷5̷ 16 18 1̷8̷ 1̷9̷ 2̷0̷ 2̷1̷ 2̷2̷ 2̷5̷

The 7th number is 16 and the 8th number is 18 so the median is 17 (halfway between 16 and 18).

Mean

To find the mean of a list of values:

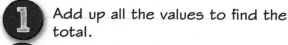 Add up all the values to find the total.

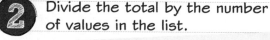 Divide the total by the number of values in the list.

Always **write down** the total first. Then divide by the number of values.

This list shows the numbers of pupils in a primary school each morning for eight days.

152 165 165 163 159 160 160 158

Total = 152 + 165 + 165 + 163 + 159 + 160 + 160 + 158 = 1282

There are 8 values in the list.

Mean = 1282 ÷ 8 = 1282 = 160.25

Now try this

tier F

1 Here is a list of the numbers of patients a doctor saw in the first 10 working days of a new job.

18 16 26 18 15 17 23 18 24 28

(a) Find (i) the mode, (ii) the median, (iii) the mean.   **(4)**

On the 11th day, the doctor saw 20 patients.

(b) Will the mean for all 11 days be greater or less than the mean for the first 10 days? Give a reason.   **(1)**

Mean from a frequency table

To find the mean from a frequency table, you can use the formula mean $= \frac{\sum fx}{\sum f}$ given on the formulae sheet in the exam.

The symbol '$\sum$' means 'sum of', so $\sum f$ means the sum of all the frequencies in the table and $\sum fx$ means the sum of all the products $f \times x$ where the x values are from the table.

Mean of discrete data from a frequency table

Use the formula: mean $= \frac{\sum fx}{\sum f}$

Add another column to the right of the table for $f \times x$.

The table here gives information about European shoe sizes of 20 people.

The mean $= 598 \div 20 = 29.9$

28, 30, 32 and 34 are the x values.

Shoe size	Frequency	$f \times x$
28	8	$8 \times 28 = 224$
30	7	$7 \times 30 = 210$
32	3	$3 \times 32 = 96$
34	2	$2 \times 34 = 68$
	20	598

These two columns are the original table.

This is $\sum f$ (the sum of the frequencies).

This is $\sum fx$.

Worked example

tier **F**

The table shows information about the numbers of letters in the first names of a group of 50 people.

Number of letters	Frequency	$f \times x$
3	2	$2 \times 3 = 6$
4	5	$5 \times 4 = 20$
5	14	$14 \times 5 = 70$
6	19	$19 \times 6 = 114$
7	10	$10 \times 7 = 70$

(a) Work out the mean number of letters in the names of the 50 people. **(2)**

$$\frac{6 + 20 + 70 + 114 + 70}{50} = \frac{280}{50} = 5.6$$

(b) Another person called Simon joins the group.
Will the mean of the first names in the group increase?
You must give a reason for your answer. **(1)**

The mean will not increase because 'Simon' has 5 letters which is less than the old mean of 5.6

Start by adding a column for $f \times x$.
Work out each $f \times x$ and write down the answer.
If $\sum f$ is not given in the question then work it out by adding the frequencies.
Use mean $= \frac{\sum fx}{\sum f}$ from the formulae sheet

EXAM ALERT!

Some students wrongly think that the answer must be a whole number. Always give your answer to an appropriate degree of accuracy.

Students have struggled with exam questions similar to this – **be prepared!**

Results P

There is no need to do any calculations. Compare the mean with the number of letters in Simon's name.

Now try this

tier **F**

1 The table gives information about the numbers of mobiles in some households.

Number of mobiles	0	1	2	3	4
Frequency	5	12	23	30	14

Calculate the mean. Give your answer correct to 1 decimal place. **(2)**

A common mistake is to write $5 \times 0 = 5$ instead of 0.
Check your working carefully.

Mean from a grouped frequency table

You can estimate the mean from a grouped frequency table by using the formula mean $= \dfrac{\sum fx}{\sum f}$ where $\sum$ stands for 'the sum of', f stands for the frequencies and x stands for the **midpoints** of the intervals.

You can use $\bar{x}$ ('x bar') as the symbol for the mean.

Mean of continuous data from a frequency table

The table gives information about the times some students spent on homework.

The midpoint of each interval is found by adding the end points and dividing by 2.

The midpoint of the interval $20 < T \leqslant 30$ is $x = \dfrac{20 + 30}{2} = 25$

Because you are using the midpoint of each interval, you are working out an **estimate** for the mean. You would need to know the time taken by every student to find the exact value.

Time, T (mins)	Frequency	Midpoint (x)	$f \times x$
$0 < T \leqslant 10$	12	5	$12 \times 5 = 60$
$10 < T \leqslant 20$	8	15	$8 \times 15 = 120$
$20 < T \leqslant 30$	3	25	$3 \times 25 = 75$
$30 < T \leqslant 40$	2	35	$2 \times 35 = 70$
	25		325

This is $\sum f$ (the sum of the frequencies).   This is $\sum fx$.

Using the formula, $\bar{x} = \dfrac{\sum fx}{\sum f} = 325 \div 25 = 13$

Worked example

tier **F&H**

The table shows information about the number of years a group of people had been driving.
Calculate an estimate for the mean. **(3)**

Number of years of driving (N)	Frequency	Midpoint (x)	$f \times x$
$0 < N \leqslant 10$	8	5	$8 \times 5 = 40$
$10 < N \leqslant 20$	12	15	$12 \times 15 = 180$
$20 < N \leqslant 30$	14	25	$14 \times 25 = 350$
$30 < N \leqslant 40$	11	35	$11 \times 35 = 385$
$40 < N \leqslant 50$	5	45	$5 \times 45 = 225$
$\sum f = 50$			$\sum fx = 1180$

$\bar{x} = 1180 \div 50 = 23.6$ years

1. Work out the midpoints.
2. Work out each $f \times x$ and find their total.
3. Add up the frequencies.
4. Then use the formula to work out the mean.

EXAM ALERT!

The question will ask you to calculate an estimate for the mean. It is an estimate because you are using midpoints. It does **not** mean that you should round your answer.

Students have struggled with exam questions similar to this – **be prepared!**

 ResultsPlus

Now try this

tier **F&H**

You need to work out the midpoint of each interval first.

1 The table gives information about the areas, A (in m^2) of plots of wasteland in a town.

Area, A (m^2)	$0 < A \leqslant 200$	$200 < A \leqslant 400$	$400 < A \leqslant 600$	$600 < A \leqslant 800$	$800 < A \leqslant 1000$
Frequency	56	40	28	20	8

(a) Calculate an estimate for the mean area of a plot of wasteland. **(3)**

(b) Explain why your answer is an estimate. **(1)**

Mode and median from a frequency table

Mode

In a frequency table for **discrete** data, the mode is the value with the **highest frequency**. Be careful – the mode is the **data value** and **not** the frequency.

Number of cars	0	1	2	3	4	5
Frequency	5	16	12	10	7	3

The mode is the number of cars with the greatest frequency – so the mode is 1.

The table gives information about the number of cars parked in the grounds of some buildings.

Median

The median is the middle value when the data is written in order of size.

In a frequency table the data is already in order.

Make an extra column headed Cumulative frequency.

Fill in the column by starting with 5.

The median is found by using the rule

$$\text{median} = \frac{n + 1}{2}$$

where n is the total frequency.

Number of cars	Frequency	Cumulative frequency
0	5	5
1	16	21
2	12	33
3	10	43
4	7	50
5	3	53

Add a cumulative frequency column

Write down the fi frequency.

5 + 16 = 21 goes here.

The final number i this column shoul equal the total frequency.

The median is the $\frac{53 + 1}{2}$th value.

The 27th value is 2. The median is 2.

Worked example

tier F&H

The table gives information about the numbers of children in some families.

Number of children	Frequency	Cumulative frequency
0	8	8
1	11	19
2	10	29
3	7	36
4	4	40

(a) Write down the mode. **(1)**

The mode is 1.

(b) Work out the median. **(2)**

From the cumulative frequency column $\frac{n + 1}{2}$ = 20.5. The median is 2.

For number of children, 1 has the largest frequency.

Find the mean of the 20th and 21st values.

Now try this

tier F&H

1 The table shows class attendance.

Attendance	22	23	24	25	26
Frequency	8	12	15	12	12

(a) Write down the mode. **(1)**
(b) Work out the median. **(2)**

Averages from grouped frequency tables

The mode and median cannot be given for grouped data because you do not know the exact data values. Instead you need to find the modal class and the interval containing the median.

Modal class and interval containing the median

The **modal class** is the interval which has the highest frequency.

Here is a table giving information about the amount of time some trains were late.

The interval that contains the median is found by using cumulative frequency.

A column for cumulative frequencies has been added to the table.

The median is the $\frac{n+1}{2}$th data value in the table where n is the total of the frequencies.

> Make sure you pick the class interval that contains the median, and **not** the middle class interval.

The modal class is $0 < T \leqslant 5$ because it has the greatest frequency.

Time late, T (minutes)	Frequency	Cumulative frequency
$0 < T \leqslant 5$	12	12
$5 < T \leqslant 10$	8	20
$10 < T \leqslant 15$	11	31
$15 < T \leqslant 20$	8	39
$20 < T \leqslant 25$	5	44
$25 < T \leqslant 30$	2	46
$30 < T \leqslant 35$	1	47

$\frac{n+1}{2} = \frac{47+1}{2} = 24$

The 24th data value lies in the interval
$10 < T \leqslant 15$

The lowest 20 times go up to here. The 21st, 22nd, ... are in the next interval.

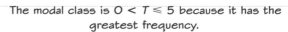

Worked example

tier **H**

The table gives information about the ages of people going on an outing.

Age, N (years)	Frequency	Cumulative frequency
$5 \leqslant N < 15$	18	18
$15 \leqslant N < 25$	20	38
$25 \leqslant N < 35$	16	54
$35 \leqslant N < 45$	21	75
$45 \leqslant N < 55$	16	91

Calculate the median. Give your answer correct to 3 significant figures. **(3)**

Median $= b + \left(\frac{\frac{1}{2}n - f}{f_c}\right) c = 25 + \left(\frac{\frac{1}{2}(91) - 38}{16}\right) 10 = 29.7$

Where
b = lower class boundary of the median class, 25
f = sum of all the frequencies below b, 38
f_c = frequency of the median class, 16
c = class width of the median class, 10

Now try this

tier **F&H**

1 The table gives information about the numbers of days some people had off work.

Days off, D	$0 \leqslant D < 5$	$5 \leqslant D < 10$	$10 \leqslant D < 15$	$15 \leqslant D < 20$	$20 \leqslant D < 25$
Frequency	12	8	11	8	5

(a) (i) Write down the modal class. (ii) Write down the class interval that contains the median. **(1)**
(b) Calculate the median. Give your answer correct to 3 significant figures. **(3)**

Which average?

You need to know the advantages and disadvantages of the three averages.

You also need to know how to interpret answers that involve averages.

Advantages and disadvantages of the three averages

	Advantages	Disadvantages
Mode	Easy to find Not affected by extreme values Can be used with non-numerical data	May not exist Cannot be used for advanced work
Median	Not affected by extreme values	Cannot be used for advanced work
Mean	Uses all the data Can be used for advanced work (e.g. calculating measures of spread)	Affected by extreme values and rarely occurs in the data

How extreme values affect the mean and median

Here is a list of the numbers of times some people went to the doctor's in one year.

2 3 3 4 4 5 6 8 9 10 30

The median is the 6th value, so 5 visits.

The mean is 84 ÷ 11 = 7.6 visits.

If we ignore the extreme value of 30,
the median changes to the 5.5th value, so 4.5 visits
and the mean changes to 54 ÷ 10 = 5.4 visits.

> The extreme value affects the mean more than it affects the median.

Worked example
tier F&H

The stem and leaf diagram shows information about the number of patients in A&E on each of 15 nights.

```
1 | 2  2  4
2 | 0  3  6  7          Key  4|3 represents
3 | 1  2  3  6  7             43 patients
4 | 0  2  3
```

(a) Find the median. **(1)**

$$\frac{n+1}{2} = \frac{16}{2} = 8 \quad \text{The 8th value is 31.}$$

On the next two nights the numbers of patients were 38 and 40.

(b) What effect does this have on the median? **(1)**

$$\frac{n+1}{2} = \frac{18}{2} = 9 \quad \text{The median increases to 32.}$$

Worked example
tier F&H

The mean of 10 tests for Jim is 17.5.
Each test is marked out of 20.
Can Jim get enough marks in his next test to bring his mean mark up to 18? **(2)**

Total so far = 10 × 17.5 = 175
Desired total = 11 × 18 = 198
198 − 175 = 23 so it's not possible.

> You can multiply the mean by the number of data values to work out the total of all the data values.

Now try this
tier F&H

1 The table shows the numbers of computers in a sample of people's houses.

Number	0	1	2	3	4	5
Frequency	4	7	9	6	4	1

(a) Work out the median. **(2)**

(b) Work out the mean. **(2)**

An additional house has 15 computers.

(c) Which of the two averages will be most affected if this house is included in the sample? **(2)**

Estimating the median

You can use a cumulative frequency diagram to find an estimate for the median of grouped data.

Using a cumulative frequency diagram to estimate the median

1 Work out the total frequency, n.

2 Calculate $\dfrac{n+1}{2}$ to find the **position** of the median.

3 Read across from this value on the vertical axis to the cumulative frequency curve.

4 Read down to the horizontal axis.

The table shows how late a bus was on each of 38 days.

Time late, T (minutes)	Frequency
$0 < T \leqslant 5$	9
$5 < T \leqslant 10$	7
$10 < T \leqslant 15$	10
$15 < T \leqslant 20$	7
$20 < T \leqslant 25$	5

This is the cumulative frequency diagram obtained from the table.

Use $\dfrac{n+1}{2} = \dfrac{38+1}{2} = 19.5$ and draw a horizontal line at 19.5 to the graph. The estimate of the median is 12 minutes.

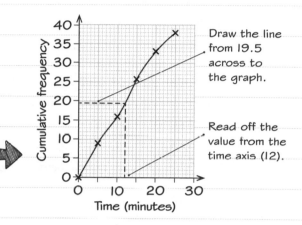

Draw the line from 19.5 across to the graph.

Read off the value from the time axis (12).

tier F&H

Worked example

The table gives information about the number of marks some students got in a test.

Marks, M	Frequency	Cumulative frequency
$0 < M \leqslant 10$	8	8
$10 < M \leqslant 20$	12	20
$20 < M \leqslant 30$	23	43
$30 < M \leqslant 40$	21	64
$40 < M \leqslant 50$	13	77
$50 < M \leqslant 60$	6	83

(a) Complete the cumulative frequency column. **(1)**
(b) Draw a cumulative frequency diagram. **(2)**
(c) Find an estimate for the median. **(2)**

$\dfrac{n+1}{2} = \dfrac{83+1}{2} = 42$ Median = 29

Draw a line from 42 on the cumulative frequency axis to the graph and read off the median.

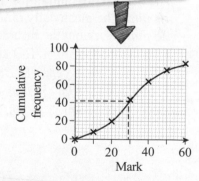

This answer of 29 is an estimate because the data has been grouped. The individual data values are unknown.

Now try this

tier F&H

1 The table gives information about the area of land, A (in hectares), held by some farms.

Area of land, A (hectares)	$0 < A \leqslant 50$	$50 < A \leqslant 100$	$100 < A \leqslant 150$	$150 < A \leqslant 200$	$200 < A \leqslant 250$	$250 < A \leqslant 300$
Frequency	3	12	26	29	23	7

By drawing a cumulative frequency diagram, find an estimate for the median. **(4)**

The mean of combined samples

You can work out the mean of combined samples if you know the mean and the size of each sample.

The mean of a combined sample

The diagrams show two samples which are combined to make a bigger-sized sample.

You can work out the mean of the combined sample by finding the total age in each of sample 1 and sample 2.

Sample 1: total age = $\bar{a_1} \times 12$

Sample 2: total age = $\bar{a_2} \times 8$

Combined sample: total age = $12\bar{a_1} + 8\bar{a_2}$

Mean = $\dfrac{\text{total age}}{\text{Total size of sample}} = \dfrac{12\bar{a_1} + 8\bar{a_2}}{12 + 8}$

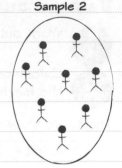

Sample 1   Sample 2

12 people, mean age = $\bar{a}_1$ 8 people, mean age = $\bar{a}_2$

Combined sample

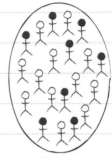

20 people

Golden rule

For samples of sizes n_1, n_2 and means x_1, x_2 the mean of the combined sample is

$$\bar{x} = \dfrac{n_1 x_1 + n_2 x_2}{n_1 + n_2}$$

Worked example

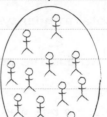
tier H

The mean daily rainfall for the first 20 days of April was 4.8 mm. The mean daily rainfall for the other 10 days was 6.6 mm. Work out the mean daily rainfall for April.   **(2)**

Total rainfall in first 20 days = 20 × 4.8 = 96

Total rainfall in next 10 days = 10 × 6.6 = 66

Total rainfall in all 30 days = 96 + 66 = 162

Mean daily rainfall = 162 ÷ 30 = 5.4 mm

EXAM ALERT!

Don't just add up the two means and divide by 2. You need to take account of the fact that the first mean was calculated over 20 days but the second mean was calculated over 10 days.

Students have struggled with exam questions similar to this – **be prepared!**

ResultsPlus

You can also use the golden rule:

$$\bar{x} = \dfrac{n_1\,\bar{x}_1 + n_2\,\bar{x}_2}{n_1 + n_2} = \dfrac{20 \times 4.8 + 10 \times 6.6}{20 + 10}$$

Now try this

tier H

1. The mean number of badges earned by the 12 younger members of a scout troop was 5.5.
   The mean number of badges earned by the 8 older members of the scout troop was 8.5.
   Work out the mean number of badges earned by all the members of the scout troop.   **(2)**

2. The mean height of a class of 25 students is 164 cm.
   The mean height of the 10 girls in the class is 161 cm.
   Work out the mean height of the boys in the class.   **(3)**

 Start by working out the total of the heights of all the students and of the girls.

Weighted means

Weighted means

A weighted mean is one where each data value is multiplied by a number (the weight) based on importance.

The weighted mean $\bar{x}$ is given by the formula

$$\bar{x} = \frac{\sum wx}{\sum w}$$

where w is the weight given to each variable, x.

For example, in an interview for a job, people have to do four tasks: A, B, C and D.

The weights given to the tasks are 1, 2, 2 and 5, meaning that task D is the most important and task A the least.

Task	A	B	C	D
Weight	1	2	2	5
Jim's mark	10	8	7	4
Anne's mark	3	4	6	8

Jim's weighted mean

$$= \frac{1 \times 10 + 2 \times 8 + 2 \times 7 + 5 \times 4}{1 + 2 + 2 + 5} = 6$$

Anne's weighted mean

$$= \frac{1 \times 3 + 2 \times 4 + 2 \times 6 + 5 \times 8}{1 + 2 + 2 + 5} = 6.3$$

Worked example

tier H

In a flower show competition, displays were given a mark for shape, a mark for colour and a mark for ambience.

The table shows the weight for each quality.

Quality	Shape	Colour	Ambience
Weight	1	2	2

Mr Smith got 7 for shape, 9 for colour and 8 for ambience.

Work out his weighted mean. **(2)**

Weighted mean $= \dfrac{1 \times 7 + 2 \times 9 + 2 \times 8}{1 + 2 + 2} = 8.2$

> 'Ambience' and 'colour' are each twice as important as shape, as they have twice the weight.

> 1. Multiply each mark by its weight then add your answers.
> 2. Divide the total by the sum of the weights.
> 3. Write down all the figures on the calculator display.

Worked example

tier H

An exam consists of Paper 1 worth 80 marks and Paper 2 worth 70 marks.

The papers are equally weighted.

Liz got 52 marks for Paper 1 and 56 marks for Paper 2.

Work out her overall percentage for the exam. **(3)**

Paper 1: $\dfrac{52}{80} \times 100 = 65\%$

Paper 2: $\dfrac{56}{70} \times 100 = 80\%$

Overall: $\dfrac{65\% + 80\%}{2} = 72.5\%$

> The first step is to work out the percentage mark on each paper. Write each mark as a fraction of the paper total and then multiply by 100.

> Since the papers are equally weighted you can add the percentages together and divide by 2 as it is the same as using weights of 1.

Now try this

tier H

1 An exam has three papers: A, B and C.
Paper A is worth 60 marks. Paper B is worth 60 marks and Paper C is worth 80 marks. The percentage marks on the papers are equally weighted.
Jimmy got 45 on Paper A, 36 on Paper B and 60 on Paper C.
What is his mean percentage? **(3)**

> You should work out the percentage mark on each paper first.

Measures of spread

Interquartile range

Range and interquartile range are measures of **spread**. They tell you how spread out the data is. **Quartiles** divide a data set into four equal parts. This page shows you how to find the quartiles of **discrete** data.

$Q_1 = \frac{n+1}{4}$ th value, where n = number of data values.

Interquartile range (IQR)

• Half of the values lie between the lower quartile and the upper quartile.

× ×   × ×× ×   × ×××   ×   DATA VALUES

Smallest value   Lower quartile (Q_1)   Median (Q_2)   Upper quartile (Q_3)   Largest value

$Q_3 = \frac{3(n+1)}{4}$ th value.

Range = largest value − smallest value

Interquartile range = (IQR) = upper quartile (Q_3) − lower quartile (Q_1)

Worked example

tier F&H

Here are the goals that Kim scored in 11 netball games.
13   10   4   10   7   12   11   14   14   8   6
(a) Work out the lower quartile.   **(1)**
4   6   ⑦   8   10   10   11   12   ⑬   14   14
Q_1 = 3rd data value = 7
(b) Work out the upper quartile.   **(1)**
Q_3 = 9th data value = 13
(c) Work out the interquartile range.   **(1)**
IQR = $Q_3 - Q_1$ = 13 − 7 = 6

1. Put the data in order of size.
2. Count the number of values.
3. Find Q_1, the $\frac{n+1}{4}$ th data value.
   In this case $\frac{11+1}{4} = 3$
4. Find Q_3, the $\frac{3(n+1)}{4}$ th data value. In this case $\frac{3(11+1)}{4} = 9$
5. Work out IQR = $Q_3 - Q_1$

Worked example

tier F&H

The table gives information about parked cars.

Number of cars	0	1	2	3	4	5
Frequency	5	16	12	10	7	3
Cumulative frequency	5	21	33	43	50	53

Find the interquartile range.   **(3)**
$Q_1 = \frac{53+1}{4}$ = 13.5th value = 1
$Q_3 = 3 \times 13.5$ = 40.5th value = 3
Interquartile range = 3 − 1 = 2

EXAM ALERT!

For any question in which you have to find the median or the interquartile range from a frequency table you **must** use cumulative frequencies.

Students have struggled with exam questions similar to this – **be prepared!**

Result Plu

1. Add and complete a cumulative frequency column.
2. Find n.
3. Find Q_1, the 13.5th data value.
4. Find Q_3, the 40.5th data value.
5. Work out IQR = $Q_3 - Q_1$

Now try this

tier F&H

1   A scientist counted the number of spots on 15 leaves of a rose bush.
   3   8   0   7   4   0   8   3   2   4   3   1   1   0   2
   Work out the lower quartile, the upper quartile and the interquartile range.   **(3)**

Box plots

You need to be able to draw and to interpret a box plot.

Box plots

Box plots are used to display the **minimum value**, the **maximum value**, the **lower quartile**, the **upper quartile** and the **median** from a distribution.

If you are asked to draw a box plot these are the five pieces of information you must put on the grid.

The lowest 25% of the values are less than or equal to the lower quartile, Q_1.

The highest 25% of the values are greater than or equal to the upper quartile, Q_3.

Make sure you do not extend this line **through** the box.

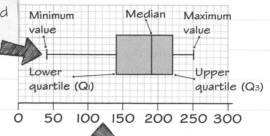

A box plot is always drawn on graph paper and always includes a scale.

Interpreting box plots

The box plot shows information about the total numbers of passengers on different bus routes one morning.

The lowest number of passengers was 40.

The median was 160.

$Q_1 = 140$ so 25% of the bus routes had 140 passengers or fewer.

$Q_3 = 220$ so 25% of the bus routes had 220 passengers or more.

The interquartile range was $Q_3 - Q_1 = 220 - 140 = 80$

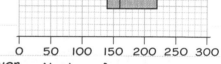

Number of passengers

Worked example

tier F&H

Here are the ages of some people in a club.

10 11 11 12 15 18 18 19 23 29
35 36 41 41 48

Draw a box plot to show this information. **(3)**

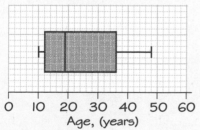

Age, (years)

The ages are already in order. If they were not, the first step would be to put them in order.

1. The lowest value is 10 and the highest value is 48.
2. Draw the scale from 0 to 60 and mark the lowest and highest values.
3. Count the number of values, n.
4. $\frac{n+1}{2} = 8$ so the median is the 8th value (19).
5. $\frac{n+1}{4} = 4$ so Q_1 is the 4th value (12).
6. $\frac{3(n+1)}{4} = 12$ so Q_3 is the 12th value (36).
7. Complete the diagram by marking Q_1, the median and Q_3.

Now try this

tier F&H

Start by putting the data values in order, smallest first.

1 The list shows some information about the numbers of people attending a community centre over 15 nights.

14 17 43 25 12 25 34 43 28 19 15 23 24 34 30

Draw a box plot to show this information. **(3)**

Interquartile range and continuous data

You can use a cumulative frequency diagram to find the quartiles and the interquartile range of grouped data. For **continuous** data:

$Q_1 = \dfrac{n}{4}$ th value

$Q_2 = \dfrac{n}{2}$ th value

$Q_3 = \dfrac{3n}{4}$ th value

The cumulative frequency diagram on the right shows the lateness of 48 trains. You can read the values for Q_1 and Q_3 off the graph.

$Q_1 = 5$ and $Q_3 = 18$

Interquartile range $= Q_3 - Q_1 = 18 - 5 = 13$ minutes

$Q_3 = \dfrac{3n}{4} = $ 36th value

Read across from 36 and down to the horizontal axis. $Q_3 = 18$ minutes

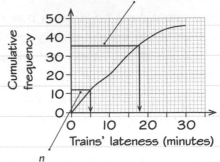

$Q_1 = \dfrac{n}{4} = $ 12th value

Read across from 12 and down to the horizontal axis. $Q_1 = 5$ minutes

---

Worked example

tier F&H

The cumulative frequency table gives data about the distances 120 people travelled to work.

Distance travelled, x (km)	Cumulative frequency
$0 < x \le 5$	34
$0 < x \le 10$	72
$0 < x \le 15$	94
$0 < x \le 20$	110
$0 < x \le 25$	120

(a) Draw a cumulative frequency diagram for this table. **(2)**
(b) Find an estimate for the interquartile range. **(2)**

$Q_1 = \dfrac{n}{4} = \dfrac{120}{4} = $ 30th value ≈ 4.5 km

$Q_3 = 3 \times 30 = $ 90th value ≈ 14 km

Interquartile range $= Q_3 - Q_1 \approx 14 - 4.5 = 9.5$ km

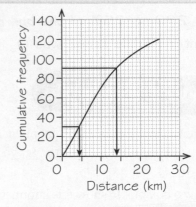

When you have to find the median or quartiles, always draw the cumulative frequency diagram — even if you are not told to do so in the exam.

Now try this

tie F&

1   This table gives information about the times, in seconds, people could read without blinking.

Time, t (seconds)	$0 < t \le 10$	$0 < t \le 20$	$0 < t \le 30$	$0 < t \le 40$	$0 < t \le 50$	$0 < t \le 60$
Cumulative frequency	4	12	23	38	52	56

Use a cumulative frequency diagram diagram to find an estimate for the interquartile range. **(4)**

Percentiles and deciles

Deciles divide data into 10 equal parts and **percentiles** divide data into 100 equal parts. You can find the deciles and percentiles of continuous data using a cumulative frequency diagram.

Finding deciles

To find the dth decile read across from $\frac{d}{10} \times n$ on the vertical axis of a cumulative frequency diagram.

Finding percentiles

To find the pth percentile read across from $\frac{p}{100} \times n$ on the vertical axis of a cumulative frequency diagram.

You may be asked to find an **interpercentile range**. For example, the 10th–90th interpercentile range is the 90th percentile minus the 10th percentile. It uses the middle 80% of the data.

This cumulative frequency diagram shows how late 50 people were for work one Monday.

To find the 76th percentile read across from $\frac{76}{100} \times 50 = 38$.
The 76th percentile is 19 minutes.

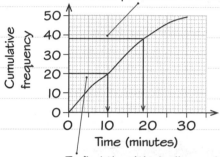

To find the 4th decile read across from $\frac{4}{10} \times 50 = 20$.
The 4th decile is 10 minutes.

Worked example

tier H

The cumulative frequency diagram gives data about the distances 150 people travelled to work.

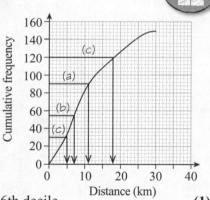

(a) Find an estimate for the 6th decile. **(1)**

6th decile: $\frac{6}{10} \times 150 = 90$th value ≈ 11 km

(b) Find an estimate for the 36th percentile. **(2)**

36th percentile: $\frac{36}{100} \times 150 = 54$th value ≈ 7 km

(c) Work out the 20th–80th interpercentile range. **(2)**

$18 - 5 = 13$ km

You can read the values off the graph. Start by working out the position up the axis by using $\frac{6}{10} \times 150$.

Work out $\frac{36}{100}$ of the total frequency, **not** $\frac{36}{100}$ of the value along the horizontal axis.

Find the 80th percentile and the 20th percentile. Work out 80th percentile – 20th percentile.

Now try this

tier H

1 The cumulative frequency table gives information about the times, in seconds, people can balance on one leg.

Time, t (seconds)	$0 < t \le 10$	$0 < t \le 20$	$0 < t \le 30$	$0 < t \le 40$	$0 < t \le 50$	$0 < t \le 60$
Cumulative frequency	5	14	24	38	58	60

(a) Draw a cumulative frequency diagram. **(2)**

(b) Use the diagram to find estimates for the value of (i) the 20th percentile, (ii) the 9th decile. **(3)**

Comparing discrete distribution

Comparing lists

Here is a list of marks for the boys and girls in Mrs Smith's class.

Boys	6	7	12	13	16	16	18	20
Girls	8	8	9	13	14	15	20	

To compare the distributions, first of all use the median or the mean.

Then work out the range for each list.

Boys' median = $\dfrac{13 + 16}{2}$ = 14.5

Girls' median = 13

In general the boys had higher marks than the girls because they had a higher median.

Boys' range = 20 − 6 = 14
Girls' range = 20 − 8 = 12

The spread of the boys' marks was higher than the girls' because the boys had a higher range.

Golden rule
When comparing data:
- always work out an average and make a comment
- always work out a measure of spread and make a comment.

Worked example

Here is a list of the number of cars arriving at a crossroads each minute for 15 minutes.

2 2 3 ⑤ 5 6 6 ⑥ 7 7 7 ⑧ 9 9 11

Information about the number of vans arriving at the crossroads each minute is shown in this box plot.

(a) Work out the median for the cars. **(1)**
Median = 6
(b) Work out the lower quartile for the cars. **(1)**
Lower quartile = 5
(c) Compare the distribution of the cars with the distribution of the vans. **(2)**
The median for the cars is 6 and for the vans is 5.
On average more cars arrived each minute.
The interquartile range for the cars is 8 − 5 = 3 and for the vans is 7 − 3 = 4 showing that the distribution of the cars is more concentrated about the median.

There are 15 numbers in the list, so the median is the $\dfrac{15 + 1}{2}$ th number.

The lower quartile is the $\dfrac{15 + 1}{4}$ th value = 4th number in the list.

Start by comparing medians from your calculation and from the box plot.

Then work out the upper quartile in the list and the interquartile range.
Compare this value with the value found from the box plot. You could also compare the ranges of the two distributions.

Now try this

1 Here is a list of the numbers of shops the people in group A visited in one week.
5 8 6 7 9 1 4 6 2 12 10 3 0 8 13
For the people in group B, the median was 7.5 and the interquartile range was 5.
Compare the two groups, A and B.

(3)

Cumulative frequency diagrams and box plots

Cumulative frequency diagrams and box plots have different uses.

A box plot provides a summary of the key values of a distribution. It can be drawn from a cumulative frequency diagram.

Drawing a box plot

A box plot can be drawn from a cumulative frequency diagram if the maximum and minimum values of the distribution are known.

This diagram shows the times taken by 43 laboratory rats to run through a maze.

The minimum time was 32 seconds.

The maximum time was 140 seconds.

$Q_1 = \dfrac{43 + 1}{4} = $ 11th value $\approx$ 51 seconds

Median = 22nd value $\approx$ 70 seconds

$Q_3 = \dfrac{3(43 + 1)}{4} = $ 33rd value $\approx$ 86 seconds

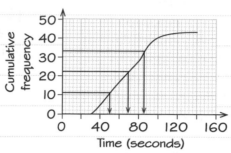

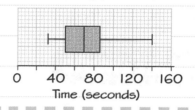

Worked example

 tier **H**

The cumulative frequency diagram gives information about the marks of 120 students in an exam.
The highest mark was 95 and the lowest mark was 12.

Draw a box plot to show this information. **(3)**

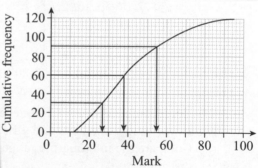

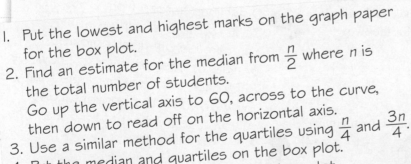

1. Put the lowest and highest marks on the graph paper for the box plot.
2. Find an estimate for the median from $\dfrac{n}{2}$ where n is the total number of students.
   Go up the vertical axis to 60, across to the curve, then down to read off on the horizontal axis.
3. Use a similar method for the quartiles using $\dfrac{n}{4}$ and $\dfrac{3n}{4}$.
4. Put the median and quartiles on the box plot.
5. Join up the five values to make the box plot.

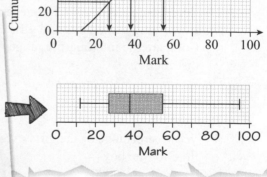

You should draw a cumulative frequency diagram first.

Now try this

 tier **H**

1 The cumulative frequency table gives information about the heights of some people.

Height, h (cm)	$h \leq 110$	$h \leq 120$	$h \leq 130$	$h \leq 140$	$h \leq 150$
Frequency	8	28	44	49	50

The minimum height was 104 cm. The maximum height was 145 cm.
Draw a box plot to show this data. **(4)**

Using cumulative frequency diagrams and box plots

Both cumulative frequency diagrams and box plots can be used to compare distributions.

Advantages and disadvantages

You can draw a box plot from a cumulative frequency diagram if you know the largest and smallest values. The box plot does not contain as much information as the diagram but it does have some advantages.

	☺	☹
Cumulative frequency diagram	Shows more information about the distribution	Harder to interpret The extreme values of the distribution are not known
Box plot	Easier to interpret Shows the extreme values	Shows less detail

For comparing distributions, it is useful to have two box plots one above the other, and to have two cumulative frequency diagrams on the same grid.

Worked example

tier H

Use the key to decide which curve to use.
Since the diagram shows percentages, the numbers of boys and girls may not be the same.

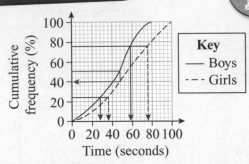

Key
— Boys
- - - Girls

The curve for the boys is to the left of that for the girls. This means that the boys' times were generally shorter.

The percentage cumulative frequency distributions show how long some boys and girls took to do a task.

(a) Compare the interquartile ranges for the two distributions. **(2)**

Boys: 58 − 28 = 30
Girls: 76 − 36 = 40
The girls had a larger IQR.

(b) What percentage of the girls exceeded the median time for the boys? **(2)**

100 − 40 = 60%

Draw lines across from 25% and 75% to the two curves.
You can use different coloured pencils or dashed lines to show the difference between the lines that you draw down to the horizontal axis.
Always use a ruler to draw the lines.

Start at 50 and draw a line to the boys' curve.
Then draw a line down to the girls' curve.
Then draw a line back to the vertical axis.

Write down the value of each quartile, as you should show your working.
Remember to answer the question, by stating that the girls' IQR is larger.

Now try this

1 The table gives information about the lengths of snakes in a zoo.

Length, l (cm)	$l \leqslant 110$	$l \leqslant 120$	$l \leqslant 130$	$l \leqslant 140$	$l \leqslant 150$
Frequency	10	35	47	53	55

The median length of snakes in the wild is 132 cm and their interquartile range is 16 cm.
Compare the distribution of snakes in the zoo with the distribution in the wild. **(4 mark**

Box plots and outliers

An extreme value in a distribution is called an **outlier**. You can use the **quartiles** and the interquartile range to define outliers.

tier H

Golden rule

An outlier is any data value that is
✓ less than $Q_1 - 1.5 \times IQR$
✓ greater than $Q_3 + 1.5 \times IQR$

Worked example

Here is some information about the test scores at a school.
    Lower quartile: 21%
    Median: 40%
    Upper quartile: 49%
Amy scored 95% on the test.
Show that her result is an outlier. **(1)**
$IQR = 49 - 21 = 28\%$
$Q_3 + 1.5 \times IQR = 49 + 1.5 \times 28 = 91\%$
$95\% > 91\%$ so Amy's result is an outlier.

Outliers on box plots

You can represent an outlier with a cross (×) on a box plot. Your 'whiskers' should only extend as far as the lowest and highest data values that are not outliers.

This data shows the numbers of times 11 friends checked social media in one day. The upper and lower quartiles are circled.

3 10 ⑭ 15 16 16 18 18 ⑲ 20 21

$IQR = Q_3 - Q_1 = 19 - 14 = 5$

The box plot on the right represents this data.

The smallest data value that is **not** an outlier is 10, so draw the end of this whisker at 10.

The highest data value is not an outlier, so draw this whisker at 21.

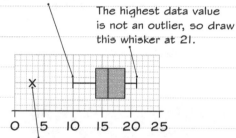

Any outliers are shown as crosses.
$Q_1 - 1.5 \times IQR = 14 - 1.5 \times 5 = 6.5$
so any data values less than 6.5 are outliers. The only outlier is 3.

Worked example

tier H

The list shows the numbers of times people accessed their email account, in one day.

0 0 1 3 3 4 5 6 7 7 9 9 10 15 19

(a) Work out the values of the quartiles. **(1)**
Q_1 is the 4th value, $Q_1 = 3$
Q_3 is the 12th value, $Q_3 = 9$
(b) Identify any values which are outliers. **(2)**
$Q_3 - Q_1 = 6$ $1.5 \times 6 = 9$
$Q_3 + 9 = 18$, so 19 is an outlier.
$Q_1 - 9 = -6$ so there are no low outliers.

The numbers are already in order. If they were not, the first step would be to put them in order.

$\dfrac{n+1}{4} = \dfrac{15+1}{4} = \dfrac{16}{4} = 4$

The IQR is $Q_3 - Q_1$. Work out $1.5 \times IQR$.

$19 > 18$ so there is an outlier at the top end. $0 > -6$ so there are no outliers at the bottom end.

Now try this

tier H

1 The list shows information about the numbers of people attending a keep-fit class over 11 days.
  4 5 8 9 10 13 16 17 17 20 24

(a) Work out the values of the quartiles. **(1)**
(b) Identify any values which are outliers. **(2)**

Box plots and skewness

The **skew** or skewness of a set of data is a description of the shape of its distribution – for GCSE that means a description of how the shape differs from a symmetrical distribution.

The diagrams show examples of the three different cases and also of the box plots that could be drawn from each of the three histograms. You can tell the skew from the box plot.

Negative skew

The longer tail is at the negative end.

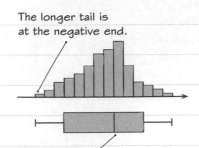

The median lies towards the right-hand end of the box plot.

Positive skew

The longer tail is at the positive end.

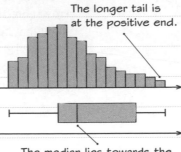

The median lies towards the left-hand end of the box plot.

Symmetrical

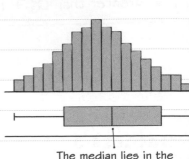

The median lies in the middle of the box plot.

A distribution has two **tails**. The 'positive' one is on the right, the 'negative' one on the left.

Worked example

The list shows the numbers of times some people used their mobiles in one day.

9 1 1 2 8 4 20 4 5 7 9 10 11 16 8

(a) Draw a box plot to show this information. **(3)**

1 1 2 ④ 4 5 8 ⑧ 7 9 9 ⑩ 11 16 20

(b) Describe and interpret the skew of the data. **(2)**

The distribution has positive skew.
The range of heavy mobile use is greater than the range of light mobile use.

> The first step is to put the numbers in order, smallest first.

tier F&H

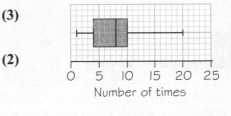

Number of times

Use $\frac{n+1}{4}$, $\frac{n+1}{2}$ and $\frac{3n+1}{4}$ to locate Q_1, the median and Q_3.

The box plot has a long tail from the upper quartile to 20.

Now try this

1 The list shows the numbers of people attending an art club for 11 days.

4 6 8 9 11 13 15 17 17 21 24

(a) Draw a box plot. **(2)**
(b) Describe and interpret the skew of the data. **(2)**

tie F&

Variance and standard deviation

The standard deviation of a distribution is a measure of its dispersion.

Definitions

The **standard deviation** of a list of numbers is given by:
$$\sqrt{\frac{\sum x^2}{n} - \left(\frac{\sum x}{n}\right)^2}$$

The **variance** of a list of numbers is the standard deviation squared.

You need to know the term 'variance' but the standard deviation is the measure used for the spread of the data.

Another expression for the standard deviation is:
$$\sqrt{\frac{\sum (x - \bar{x})^2}{n}}$$

This is not so useful for calculations.

The closer the xs are to $\bar{x}$, the smaller $\sum (x - \bar{x})^2$ is, and the smaller the standard deviation.

Finding the standard deviation

To find the standard deviation of a list of values, for example these:

70   100   130   80   90   40   60   70

Mean $= \dfrac{\sum x}{n} = \dfrac{640}{8} = 80$ — Work out the mean.

$\sum x^2 = 70^2 + 100^2 + 130^2 + 80^2 + 90^2$
$\qquad + 40^2 + 60^2 + 70^2 = 56\,400$

Square every value and find the total.

Standard deviation

$= \sqrt{\dfrac{56\,400}{8} - 80^2} = \sqrt{650}$ — Use the formula.

$= 25.5$ correct to 3 significant figures.

Advantages and disadvantages of the three measures of spread

Range	Interquartile range	Standard deviation
✓ Easy to find	✓ Not affected by extreme values	✓ Used for advanced work
✗ Affected by extreme values	✗ Difficult to calculate	✓ Uses all the data
✗ No use for advanced work	✗ Only uses the central 50% of the distribution	✗ Affected by extreme values
		✗ Harder to calculate

Worked example

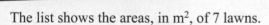 tier **H**

The list shows the areas, in m², of 7 lawns.

14 15 18 20 21 24 28

Work out the standard deviation of these areas. **(2)**

Mean $= \frac{1}{7}(14 + 15 + 18 + 20 + 21 + 24 + 28)$

$\qquad = 20$ m²

$\sum x^2 = 14^2 + 15^2 + 18^2 + 20^2 + 21^2$
$\qquad\quad + 24^2 + 28^2 = 2946$

Standard deviation $= \sqrt{\dfrac{2946}{7} - 20^2} = 4.57$ m²

Work out the mean first.
Square each value and find the sum of the squares.

The units of the standard deviation are the same as those of the data.

Now try this

 tier **H**

1 The number of visits, x, to a dentist was recorded over 10 days.
$\sum x = 200$, $\sum x^2 = 480$
Work out the mean, the variance and the standard deviation. **(3)**

Standard deviation from frequency tables

For grouped data the formula for the standard deviation is: $\sqrt{\dfrac{\sum fx^2}{\sum f} - \left(\dfrac{\sum fx}{\sum f}\right)^2}$ ← This is the mean of the data values.

If the data is continuous then the formula gives an estimate for the standard deviation – just as the formula on page 33 for the mean gave an estimate for the mean of continuous data.

The table gives information about the time (in seconds) it took some students to solve a puzzle.

The three columns in red are working columns you should add to the table to find the values you will need to calculate the standard deviation:

Time, T (s)	Frequency f	Mid T	fT	fT^2
$0 < T \leqslant 20$	6	10	60	600
$20 < T \leqslant 40$	11	30	330	9 900
$40 < T \leqslant 60$	13	50	650	32 500
$60 < T \leqslant 80$	6	70	420	29 400
$80 < T \leqslant 100$	4	90	360	32 400
	40		1820	104 800

$\sum f = 40$ $\sum fT = 1820$ $\sum fT^2 = 104\,800$

Mean = 1820 ÷ 40 = 45.5

Standard deviation = $\sqrt{104\,800 \div 40 - 45.5^2} = 23.4$ to 3 s.f.

In most questions you will not need to calcula[te] every value in a table.

Worked example

tier H   Aiming higher

The partially completed table gives data about the distances that some birds flew from their nests and back.

Distance, D (m)	Frequency	Mid D	fD	fD^2
$0 < D \leqslant 20$	19	10	190	1900
$20 < D \leqslant 40$	14	30	420	12 600
$40 < D \leqslant 60$	10	50	500	25 000
$60 < D \leqslant 80$	10	70	700	49 000
$80 < D \leqslant 100$	7	90	630	56 700
	60		2440	145 200

fD^2 means $f \times D^2$ so $14 \times 30^2 = 14 \times 90[0]$

(a) Complete the table.   (1)
(b) Estimate the standard deviation.   (2)

Put the sums of the f, the fD and the fD^2 columns at the bottom of the table.

$\sum f = 60$ $\sum fD = 2440$ $\sum fD^2 = 145\,200$

Mean = 2440 ÷ 60 = 40.66667 m

Standard deviation = $\sqrt{145\,200 \div 60 - 40.66667^2}$
                   = 27.7 (3 s.f.)

Work out the mean. Then use the formula for standard deviation.

Now try this

tier H   Aiming higher

1 The times, t seconds, that a group of 30 people took to solve a puzzle are summarised by the following data: $\sum ft = 1200$, $\sum ft^2 = 51\,000$.
   Find estimates for the mean and the standard deviation.   (3)

2 The number of children, x, seen at a surgery on 20 days has the following data:
   $\bar{x} = 10$, $\sum fx^2 = 2320$
   (a) Work out the standard deviation.   (2)
   On each of days 21 and 22, 10 children were seen at the surgery.
   (b) Compare the mean and the standard deviation for all 22 days with the mean and the standard deviation for the first 20 days.   (2)

Simple index numbers

Simple index numbers are a way of tracking changes in value through time.

Calculating an index number

Index numbers are like percentages which describe changes in costs or prices from year to year.

The cost or prices are compared with a **base year**.

In the **base year** the index number is defined as 100.

You can calculate a simple index number for year n using the rule: $\text{index} = \dfrac{\text{cost in year } n}{\text{cost in base year}} \times 100$

An index number more than 100 means that the price in year n is higher than in the base year.

If a price goes down then the index number will be less than 100.

2010

The index number of the laptop cost in 2010 (the base year) is defined as 100.

2012

In 2012, the index number is $\dfrac{425}{500} \times 100 = 85$.

Using an index number

You can use index numbers to work out costs and prices.

Use the formula:

$\text{cost in year } n = \text{cost in base year} \times \dfrac{\text{new index number}}{100}$

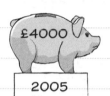

2005   **2008**

The value of an investment in 2005 (the base year) was £4000 and the index number for 2008 was 108.

So the value of the investment in 2008 was $£4000 \times \dfrac{108}{100} = £4320$.

Worked example

tier F&H

In calculations involving index numbers you will be told which year to use as the base year in the question.

The cost of some groceries in June 2012 was £120.
The cost of the same groceries in July 2012 was £121.80.
June 2012 is the base month.

(a) Work out the index number for the cost of the groceries in July 2012. **(2)**

July index number $= \dfrac{121.80}{120} \times 100 = 101.5$

Remember that when you have to **find** an index number you use $\dfrac{\text{cost in year } n}{\text{cost in base year}} \times 100$

The index number for the cost of the same groceries in August 2012 is 98.

(b) Work out the cost of the groceries in August 2012. **(2)**

Cost in August $= 120 \times \dfrac{98}{100} = £117.60$

But when you have to **use** an index number you need $\text{cost in base year} \times \dfrac{\text{new index number}}{100}$

Now try this

tier F&H

1. In 1999, the national minimum wage for an adult was £3.60 per hour.
   In 2012, it was £6.19 per hour.
   (a) Using 1999 as the base year, work out the index number for the adult minimum wage in 2012. **(2)**

   In 1999, the national minimum wage for young workers was £3.00 per hour.
   In 2012, the index number for the national minimum wage for young workers was 166.
   (b) Work out the national minimum wage for young workers in 2012. **(2)**

Chain base index numbers

A chain base index number is a comparison of this year's value with last year's value.

Finding a chain base index number

A **chain base** index number compares **this** year's value with **last** year's value (the **base**).

The calculation is very similar to finding a **simple** index number except that the base changes every year.

Chain base index number = $\dfrac{\text{value this year}}{\text{value last year}} \times 100$

The chart shows the value of an investment over four years.

2010	£1000
2011	£1400
2012	£1750
2013	£1595

 The base year for 2012 is 2011.

Chain base index number for 2012 is $\dfrac{1750}{1400} \times 100 = 1?$

Chain base index number for 2013 is $\dfrac{1595}{1750} \times 100 = 9?$

Using a chain base index number

You can use chain base index numbers to find a new price.

Price this year =

$\dfrac{\text{chain base index number for this year}}{100} \times$ price last year

If this year's number chain base index is 120, then this year's price is $\dfrac{120}{100} \times$ last year's price.

This means the price is 20% higher than last year.

Chain base vs simple index numbers

Chain base and simple index numbers are always worked out from

$\dfrac{\text{value this year}}{\text{value in the base year}} \times 100$

- For **simple**, the base year does not change.
- For **chain base**, the base year is always the previous year.

Worked example

tier **H** Aiming higher

The table shows the price ($) of 1 ounce of gold at the start of each year.

Year	2011	2012	2013
Price ($)	1347	1498	1657

(a) Work out the chain base index numbers for 2012 and 2013. **(2)**

2012: $\dfrac{1498}{1347} \times 100 = 111.2$

2013: $\dfrac{1657}{1498} \times 100 = 110.6$

(b) The chain base index number for 2014 was 120. Work out the price of 1 ounce of gold in 2014. **(2)**

2014: $\dfrac{120}{100} \times 1657 = \$1988.4$

The denominator changes every year for the chain base. The denominator for 2012 is the 2011 price.

For 2013, you must use the 2012 price as the base.

 Use $\dfrac{\text{chain base index number for 2014}}{100} \times 2013$ price

For 2014, the price last year is the 2013 price.

Now try this

1 The table shows the price of a model of car for three years.

Year	2011	2012	2013
Price (£)	18 500	18 900	19 300

(a) Using 2011 as the base year, find the chain base index numbers for 2012 and 2013. **(3)**

(b) Explain why the two index numbers are not the same, even though the increase in each year is the same. **(1)**

Weighted index numbers

You sometimes need to measure how a particular set of items changes in value. You can do this using a **weighted index number**. Each item is assigned a different **weight** in the calculation to show how important it is.

The most well-known weighted index numbers are the Consumer Prices Index (CPI) and the Retail Price Index (RPI), which are both measures of inflation in prices.

Calculating weighted index numbers

You can calculate the weighted index number for a set of items with weights w by using the rule:

$$\frac{\sum(w \times \text{index number for each item})}{\sum w}$$

The example shows a calculation of the weighted index number for animal feed with 2010 as the base year.

The weights here reflect the ratio of barley to wheat. Twice as much barley is used as wheat.

The manufacturers of an animal feed use two tonnes of barley to one tonne of wheat.

	2010	2012
Barley (£/tonne)	100	120
Wheat (£/tonne)	128	176

The base year is 2010.

The index number for barley is $\frac{120}{100} \times 100 = 120$

The index number for wheat is $\frac{176}{128} \times 100 = 137.5$

The weights are 2 for barley and 1 for wheat.
The weighted index for this animal feed in 2012 is

$$\frac{2 \times 120 + 1 \times 137.5}{2 + 1} = 125.8$$

Worked example

tier **H**

The weighted index number will always lie between the largest and the smallest simple index numbers.
This is useful as a check on your answer.

Alec invests in shares. In 2010, his investment was 1000 shares in company A and 2500 shares in company B. He still had these investments in 2013.

	2010	2013
Company A	£2.50	£3.00
Company B	£2.80	£4.00

Work out the weighted index number for Alec's shares in 2013. (3)

Company A index number = $\frac{3.00}{2.50} \times 100 = 120$

Company B index number = $\frac{4.00}{2.80} \times 100 = 142.9$

The weights are 1000 for A and 2500 for B.

Weighted index number = $\frac{1000 \times 120 + 2500 \times 142.9}{1000 + 2500} = 136.4$

1. Work out the simple index number for each item.
2. Decide on the weights.
3. Use the rule:
$$\frac{\sum(w \times \text{index number for each item})}{\sum w}$$
to work out the weighted index.

Now try this

tier **H**   Aiming higher

1 The table gives information about the profits per item from two factories in 2011 and 2012.

	2011	2012
Factory A	£30	£32
Factory B	£50	£53

You could give the simple index for factory A a weight of 3.

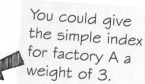

Three times as many items come from Factory A as from Factory B.
Using 2011 as the base year, work out the weighted index number for 2012. (3)

Standardised scores

Standardised scores are used to compare values from different frequency distributions. They work best when the distribution of values is symmetrical.

Understanding standardised scores

When students do different tests it is not always fair to compare the marks, as one test might have been more difficult than another.

You can use the rule:

$$\text{standardised score} = \frac{\text{mark} - \text{mean}}{\text{standard deviation}}$$

to compare marks on different tests.

A mark **greater** than the mean gives a **positive** standardised score.

A mark **lower** than the mean gives a **negative** standardised score.

The **higher** the standardised score, the **better** the student's performance on the test.

The standardised score for a mark of 80 in English is lower than for 80 in Maths, showing that the performance in Maths was better.

Bigger standard deviations result in lower standardised scores.

The diagrams show the distributions of marks in two tests.

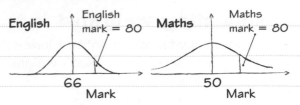

The mean mark in English was 66, and the standard deviation was 16.

The mean mark in Maths was 50, and the standard deviation was 30.

The marks were more spread out in Maths than in English.

A mark of 80 in English gives a standardised score of $\frac{80 - 66}{16} = 0.875$

A mark of 80 in Maths gives a standardised score of $\frac{80 - 50}{30} = 1.0$

Worked example

tier **H**   Aiming higher

50 people took an exam. The mark (x) for each person was recorded.

$\Sigma x = 3100$, $\Sigma x^2 = 203\,450$

Maisie got a mark of 86. Work out her standardised score. **(4)**

Mean $= \dfrac{3100}{50} = 62$

Variance $= \dfrac{203\,450}{50} - 62^2 = 225$

Standard deviation $= \sqrt{225} = 15$

Maisie's standardised score $= \dfrac{86 - 62}{15} = 1.6$

You must work out the mean mark and the variance first.
You looked at those on page 49.

1. Work out the mean using $\bar{x} = \dfrac{\Sigma x}{50}$
2. Work out the variance using
$$\frac{\Sigma x^2}{50} - \bar{x}^2$$
3. The standard deviation is the square root of the variance.

Use the formula for standardised score:
$$\frac{\text{Maisie's mark} - \text{mean}}{\text{standard deviation}}$$

Now try this

tier **H**   Aiming higher

1 Abdul took two tests in an interview for a job. On the aptitude test he got 48 marks. On the skills test he got 64 marks. The table gives information about the two tests.

	Mean mark	Standard deviation
Aptitude test	45	10
Skills test	55	8

Work out Abdul's standardised scores on the two tests. **(3)**

2 100 people took a test. The variable x denotes the mark they each got.
$\Sigma x = 4600$, $\Sigma x^2 = 226\,000$

(a) Shola got 40 marks. Work out his standardised score. **(3)**

(b) Alice had a standardised score of 1.5. Work out how many marks she got. **(2)**

Scatter diagrams and correlation

You can look for a relationship between two variables by plotting a scatter diagram.

Positive correlation	**Negative correlation**	**No correlation**

Test result
Hours of revision

Test result
Hours of TV watched

Test result
Number of bracelets worn

As one variable increases, so does the other.

As one variable increases, the other decreases.

There is no relationship between the two variables.

Types of variable

Here is a scatter diagram that shows how plants grew when given different amounts of water each day. The scatter diagram shows the heights after 14 days. There is a positive correlation.

The height of a plant is an example of a response or dependent variable.

Each point corresponds to the amount of water and the height for one plant.

Height

Amount of water

The amount of water is an example of an explanatory or independent variable.

Worked example

tier F

The table gives information about the average monthly temperature, x (°C), and the money made from coat sales, y (£000s), in a shop.

Temperature, x (°C)	4	3	8	12	18	25	22
Money made, y (£1000s)	46	50	49	23	14	4	5

(a) Plot a scatter diagram to show this information. **(2)**
(b) Describe the relationship between temperature and money made. **(1)**

Negative correlation – as the temperature increases, the money made from coat sales decreases.

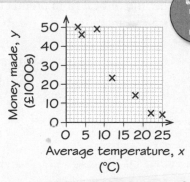

Write down the type of correlation, and answer the question in context – make sure you say how temperature affects the amount of money made.

Now try this

tier F

1 For each of the following, identify the explanatory variable and the response variable, and state what type of correlation you expect.
  (a) The weight of a child and his or her age. **(2)**
  (b) The length of a train journey and the time it takes. **(2)**
  (c) The average daily temperature and the number of cups of hot tea drunk. **(2)**
  (d) The number of bushes in a garden and the number of people who live in the house. **(2)**

Lines of best fit

You can use a line of best fit to summarise the relationship shown on a scatter diagram.
You can use your line of best fit to predict values.

Drawing the line of best fit

You can summarise data on a scatter diagram by drawing a line of best fit.

This scatter diagram shows students' estimates of the lengths of some lines whose true lengths were known by their teacher.

The line of best fit should extend before the first point and after the last point, so it covers all of the data.

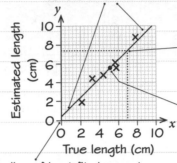

The line can be used to read off an estimate of one variable, given the value of the other. This line shows that a true length of 7 cm was, on average estimated as 7.3 cm.

This point is the mean for both x and y. If these means are given in a question, draw the line through this point. The co-ordinates are given as $(\bar{x}, \bar{y})$.

The line of best fit does not need to pass through (0, 0).

Worked example

The scatter diagram gives some information about the prices (£y) of some cars and their ages (x years).
$\bar{x} = 4$, $\bar{y} = 4600$.

(a) Draw a line of best fit on the diagram. **(1)**

(b) Use the line of best fit to predict the price of a car that is $5\frac{1}{2}$ years old. **(1)**

Price = £3400

Read off the line of best fit at $x = 5.5$

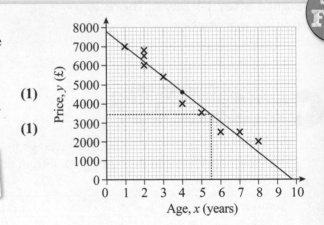

Now try this

1 The scatter diagram shows information about arm length and leg length.

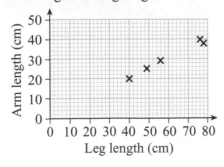

The mean leg length is 59.8 cm and the mean arm length is 30.4 cm.

(a) Draw a line of best fit. **(1)**

John has a leg length of 65 cm.

(b) Estimate his arm length. **(1)**

First, plot (59.8, 30.4) on the diagram.

Interpolation and extrapolation

You can use a line of best fit to estimate values within the range of the given data. This is **interpolation**.

You can also use it to predict values beyond the range of the given data. This is called **extrapolation**. Extrapolation is less reliable than interpolation.

The line of best fit passes through the mean point.

Using the line of best fit, a **prediction** of the height of a plant aged 3 weeks has been found by extrapolation. It is outside the known range of 5 to 25 weeks and so it is unreliable.

Using the line of best fit, an **estimate** of the height of a plant aged 13 weeks has been found by interpolation. It is between the known data of 5 weeks and 25 weeks and so it is reliable.

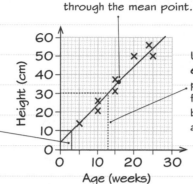

Worked example

Calculate the mean point of the data first. The line of best fit must go through the mean point.

tier F&H

The scatter diagram gives information about the heights and lifetimes of six candles.
One further candle has an initial height of 7 cm and a lifetime of 12 hours.
(a) Plot this point on the diagram.   **(1)**
(b) Using the mean point, draw the line of best fit.   **(2)**
For all seven candles the mean height is 9 cm and the mean lifetime is 15 hours.
A candle has an initial height of 15 cm.
(c) Estimate its lifetime.   **(1)**
Using the line of best fit, the lifetime is 21.5 hours.
Another candle has an initial height of 22 cm.
(d) Predict its lifetime.   **(1)**
Using the line of best fit, the lifetime is 29 hours.

This is **interpolation**. It is reliable as it uses a value within the range of the measurements.

This is **extrapolation**. It is **not** reliable as it uses a value outside the range of the measurements.

Now try this

tier F&H

1 The diagram shows information about the price paid for a second-hand car and the age of the car. The line of best fit has been drawn.

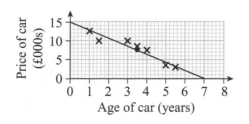

A second-hand car is sold when it is two years old.
(a) (i) Estimate its price.
    (ii) Comment on the reliability of your estimate.   **(2)**
Another second-hand car is sold when it is 6.5 years old.
(b) (i) Predict its price.
    (ii) Comment on the reliability of your estimate.   **(2)**

The equation of the line of best fit

You need to be able to find the equation of the line of best fit in the form $y = ax + b$.

The scatter diagram gives some information about the cost of a fruit cake and the percentage of fruit in it.

The equation is given in the form $y = ax + b$.

a is the gradient of the line and b is the intercept on the y-axis.

To find the gradient, draw a right-angled triangle and find its base and height.

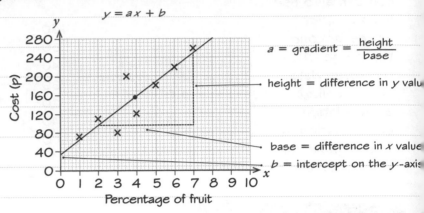

$y = ax + b$

a = gradient = $\dfrac{\text{height}}{\text{base}}$

height = difference in y valu

base = difference in x value

b = intercept on the y-axis

Worked example

tier H

The scatter diagram gives information about the temperature, $x\,°C$, and the number of sales of drinks, y, from a shop each day for 10 days.

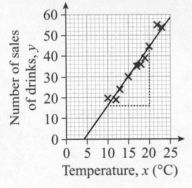

(a) Find the equation of the line of best fit. **(3)**

a = gradient = $\dfrac{\text{height}}{\text{base}} = \dfrac{45 - 16}{20 - 10} = 2.9$

$b = y - ax = 16 - 2.9 \times 10 = -13$

Equation of the line of best fit is $y = 2.9x - 13$

(b) What information about the drink sales is given by the gradient of the line of best fit? **(1)**

With each temperature increase of $1°C$, 2.9 more drinks were sold.

When choosing which x values to use, draw a large triangle with its vertices on grid lines. This will make it easier to read off the values.

When finding the base and the height, use the scale. Counting squares may give the wrong answer.

You might not be able to read b directly from the graph. If so, find a, then work out b from $b = y - ax$. Here (10, 16) was used.

Don't forget to put the values of a and b you have found into the format $y = ax + b$ to give the equation of the line of best fit.

Think about what the gradient means in the context of the question.

Now try this

ti

Remember, the correlation is negative so the gradient will also be negative.

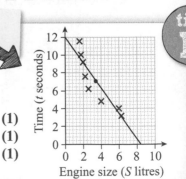

1 The scatter diagram gives information about the size (S litres) of a car engine and the time (t seconds) it takes the car to accelerate to 30 mph.
A line of best fit has been drawn on the scatter graph.

(a) Find the equation of the line of best fit. **(1)**

(b) What information is given by the gradient? **(1)**

(c) Why might the intercept in the equation not be realistic? **(1)**

Curves of best fit

If two variables have a **non-linear** relationship, you can sometimes draw a **curve** of best fit. You need to recognise the shapes of these four types of curve:

$$y = ax^n + b \qquad y = \frac{a}{x} + b \qquad y = a\sqrt{x} + b \qquad y = ka^x$$

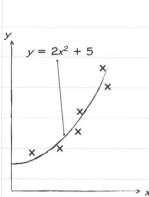

$y = 2x^2 + 5$

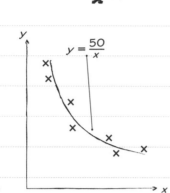

$y = \frac{50}{x}$

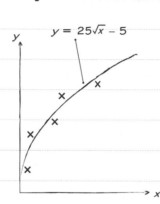

$y = 25\sqrt{x} - 5$

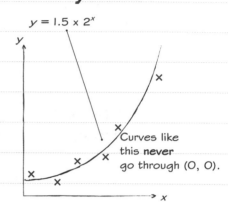

$y = 1.5 \times 2^x$

Curves like this **never** go through (0, 0).

Worked example

tier H

When a current, I amps, is passed through a resistor, the heat generated is shown by the points on the graph.

(a) Show that the curve with equation $y = 288x^2$ gives a good fit to the points. **(1)**

(b) Use the curve of best fit to estimate the heat produced when a current of 0.75 amps is passed through the resistor. Show your working by drawing on the diagram. **(1)**

160 Cals

(c) Use the equation of the curve to estimate the heat produced when a current of 3 amps is passed through the resistor. **(1)**

$y = 288 \times 3^2 = 2592$ Cals

If you need to use a curve of best fit in your exam you will always be given the equation.

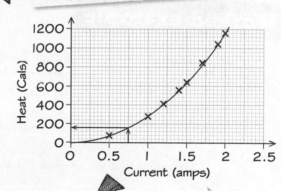

The best way to show that the curve is a good fit is to plot the curve on the graph. Use a table of values to plot points and then join them with a smooth curve.

It is easier to draw a smooth curve if you have your graph paper so your hand is **inside** the curve.

Now try this

tier H

1 The table gives the temperature, $T°C$, at different distances, d m, from a heat source.

d (m)	5	10	15	20	25	30
T (°C)	42	28	25	24	23	22

(a) Show that the curve $T = \frac{100}{d} + 20$ gives a good fit to the data. **(2)**

(b) Estimate the temperature at a distance of 8 metres from the heat source. **(1)**

Spearman's rank correlation coefficient 1

Spearman's rank correlation coefficient is used to judge the correlation of two variables. The rank correlation coefficient lies between –1 and 1 inclusive.

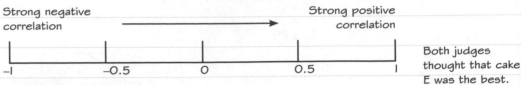

Strong negative correlation ———————→ Strong positive correlation

–1   –0.5   0   0.5   1

Both judges thought that cake E was the best.

How to calculate Spearman's rank correlation coefficient

Use the formula $1 - \dfrac{6\sum d^2}{n(n^2 - 1)}$ to work out the rank correlation coefficient.

d is the difference between the ranks for each cake.

n is the number of pairs of ranks.

d is the difference between the ranks.

The table on the right shows the ranks (positions) two judges gave to six competitors in a cake-making competition.

Competitor	A	B	C	D	E	F
Judge X	3	4	5	6	1	2
Judge Y	4	2	6	5	1	3
d	1	2	1	1	0	1

Here, $n = 6$

$\sum d^2 = 1^2 + 2^2 + 1^2 + 1^2 + 0^2 + 1^2 = 8$

Rank correlation coefficient $= 1 - \dfrac{6 \times 8}{6 \times (6^2 - 1)} = 0.7$

0.771 shows a strong positive correlation between the two judges.

tier **H**

Worked example

The table shows the positions of eight football teams ranked by points total and by goals conceded.

(a) Calculate Spearman's rank correlation coefficient for these data. **(3)**

Position by goals conceded	7	6	8	4	3	1	2	5
Position by points	1	2	3	4	5	6	7	8
d	6	4	5	0	2	5	5	3

The differences d are worked out below the table.

$\sum d^2 = 36 + 16 + 25 + 0 + 4 + 25 + 25 + 9$
$= 140$

Show your working.

$n = 8$

$1 - \dfrac{6\sum d^2}{n(n^2 - 1)} = 1 - \dfrac{6 \times 140}{8 \times (64 - 1)} = -0.667$

Be very careful to use the formula correctly.

(b) Does the table support the hypothesis, 'The higher the position in goals conceded, the lower the position in points'? **(2)**

The value –0.667 indicates a strong negative correlation between the two measures and supports the hypothesis.

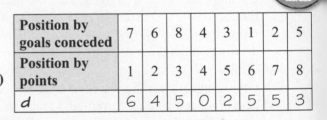

The negative number indicates that the higher one value is, the lower the other is. Make sure you make a comment about the numerical answer.

Now try this

1 The table shows the ranks for goals scored and for goals conceded by teams in a hockey league.

Goals scored	1	4	2	3	8	5	6	10	7	9
Goals conceded	10	9	6	2	8	7	5	3	4	1

(a) Calculate Spearman's rank correlation coefficient for these data. **(3)**

(b) Interpret your answer to part (a). **(2)**

Spearman's rank correlation coefficient 2

You might need to rank the data values before you can calculate Spearman's correlation coefficient.

Ranking data

The table shows the position in the league and the number of games that each of ten teams lost.

To convert the number of games lost into a rank, add an extra row labelled Rank and write numbers in order of the number of games lost.

d is then the difference between each team's league position and your rank.

League position	1	2	3	4	5	6	7	8	9	10
Number of games lost	2	9	13	16	14	11	19	18	20	26
Rank	1	2	4	6	5	3	8	7	9	10
D	0	0	1	2	0	3	1	1	0	0

Then you can work out the rank correlation coefficient using

$$1 - \frac{6\sum d^2}{n(n^2 - 1)}$$

Its value is 0.90 – a very strong positive correlation between league position and number of games lost.

Worked example

tier H

The table gives information about the age of tyres on a car and the minimum stopping distance at 50 mph.
Work out Spearman's rank correlation coefficient. **(3)**

Car	Age of tyres (months)	Stopping distance (m)	d	d²
A	10 (1)	42 (1)	0	0
B	11 (2)	49 (3)	1	1
C	15 (3)	47 (2)	1	1
D	20 (4)	53 (6)	2	4
E	24 (5)	51 (5)	0	0
F	28 (6)	58 (7)	1	1
G	30 (7)	50 (4)	3	9
H	32 (8)	60 (8)	0	0
				16

You can write the ranks beside the values in the table. The brackets are not essential. Sometimes there are separate columns for you to put them in.

Spearman's rank correlation coefficient
$$= 1 - \frac{6 \times 16}{8 \times (8^2 - 1)} = 0.81$$

Be very careful to use the formula correctly.

You can add an extra row to add up your values for d^2.

Now try this

tier H

1 The table gives information about the length and the weight of eight large birds.

Length (cm)	28	40	48	60	63	78	98	120
Weight (kg)	0.4	0.7	0.6	1.1	1.0	2.2	4.1	5.6

(a) Calculate Spearman's rank correlation coefficient. **(3)**

(b) Interpret your answer to part (a). **(2)**

Time series

Time series are graphs which show variation over time.

They are useful for studying the **trend** and the **seasonal variation**. The trend is the overall behaviour over time.

This time series shows car sales. The trend is increasing and the seasonal variation is six monthly.

A **trend line** shows you the trend of the data over time. It ignores any seasonal variation.

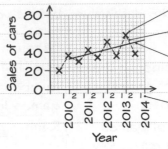

Use crosses for the points.

Join the crosses with dotted lines

Draw a **trend line** by eye to see the trend better. You can use it to predict future values.

These are the 1st and 2nd halves of each year.

For every year, the number of sales was higher in the second half of the year. This is called **seasonal variation**.

Worked example

tier **F**

The graph shows information about the number of moths caught each season in 2011 and 2012. The table gives similar information for 2013.

	2013			
Quarter	1	2	3	4
Number caught	17	24	36	14

(a) Add this information to the time series graph. **(2)**

(b) When was the greatest number of moths caught? **(1)**

In quarter 3 of 2011.

(c) Describe the seasonal variation. **(1)**

There is a four-quarterly seasonal variation with higher numbers of moths caught in quarter 3 and lower in quarters 4 and 1.

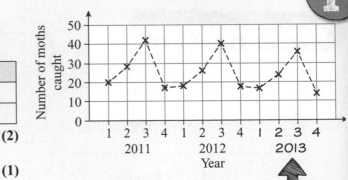

Look at the graph and see how any regular pattern varies.

Always answer in context.

Now try this

tier **F**

1 The table gives information about the number of wild flowers in a park.

	2012				**2013**			
Quarter	1	2	3	4	1	2	3	4
Number	14	25	34	23	8	23	37	20

(a) Plot this information on a time series graph. **(2)**

(b) Describe the trend in the number of wild flowers over the two years. **(1)**

(c) Write down the quarter with the highest number of wild flowers. **(1)**

Moving averages

Moving averages are a way of smoothing out the data in time series.

This makes it easier to account for seasonal variation, and to tell what the trend is.

The table gives some data about the number of voles on an island.

Year	Population (thousands)	3-point moving average (thousands)
2008	4.5	
2009	5.2	5.50
2010	6.8	5.57
2011	4.7	5.67
2012	5.5	

The moving averages do not vary as much as the original figures.

The first 3-point moving average is the mean of the first three consecutive values:

$$\frac{4.5 + 5.2 + 6.8}{3} = 5.50$$

The next 3-point moving average is the mean of the 2nd, 3rd and 4th values:

$$\frac{5.2 + 6.8 + 4.7}{3} = 5.57$$

Worked example

tier F&H

The table gives information about the number of fish caught from a lake.

Year	Quarter	Number caught (1000s)	4-point moving averages (1000s)
2011	1	20	
	2	28	
			26.75
	3	42	
			26.25
	4	17	
			25.75
2012	1	18	
			25.25
	2	26	
			25.50
	3	40	
	4	18	

Find the first 4-point moving average by finding the mean of the first four values:

$$\frac{20 + 28 + 42 + 17}{4} = 26.75$$

Make sure you put the moving averages in the correct cells – the mid-point of the values they cover. The first 4-point moving average is placed between the 2rd and 3rd values to show that it is calculated from the first four values.

Some of the moving averages can't be calculated as there are not enough data points.

(a) Complete the 4-point moving average column. **(2)**

(b) Describe the trend in the numbers caught. **(1)**

Looking at the moving average column, the number of fish caught is decreasing.

Now try this

tier F&H

	2010		2011		2012		2013		2014
Season	June	Dec	June	Dec	June	Dec	June	Dec	June
Number	20	35	30	42	33	53	37	57	38

The table gives information about the sales of cars during June and December each year.

(a) Work out the 2-point moving averages for this information. **(2)**

(b) Comment on any trend in the sales. **(1)**

Moving averages and trend lines

You can draw a trend line by plotting the moving averages on a time series graph.
The position where they are plotted depends on how many values are covered.

3-point moving averages

These moving averages show an increasing trend.

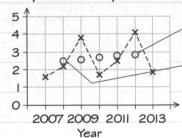

The moving averages are always plotted as circles. The circles show how the data has been smoothed out, which means you can draw a trend line.

The 3-point moving average is plotted against the middle point. So the average for 2007, 2008 and 2009 is plotted at 2008.

2-point or 4-point moving averages

2-point moving averages are plotted halfway between the 1st and 2nd points.

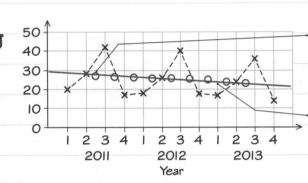

The 4-point moving averages are plotted halfway between the 2nd and 3rd points. So the 1st moving average (for quarters 1, 2, 3 and 4) is plotted at position 2.5. The next is at position 3.5.

To plot the trend line, draw the line of best fit through the moving averages.

Worked example

tier **H**

This graph shows sales of smoothies from a café.

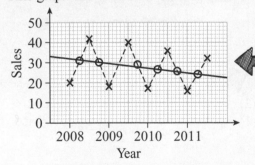

The 2-point moving averages are plotted halfway between the crosses.

(a) Plot these calculated 2-point moving averages on the time series. **(2)**

| 31 | 30 | 29 | 28.5 | 26.5 | 26 | 24 |

(b) Draw a trend line. **(1)**

(c) Describe the trend of the sales of smoothies over the years 2008 to 2011. **(1)**

The trend line shows that the sales are decreasing.

Answer the question in context here by referring to 'sales'.

Now try this

tie **H**

1		2011			2012			2013	
Rainfall (cm)	102	156	142	106	157	135	110	169	
3-point moving average		133	135	135	133	134	138		

(a) Plot the time series. **(2)**   (c) Draw the trend line. **(1)**

(b) Plot the moving averages. **(2)**   (d) Describe the trend. **(1)**

The equation of the trend line

You need to be able to find the equation of a trend line you have drawn on a time series graph.

Equation of the trend line

Plot the moving averages and draw the line of best fit. Then you can write an equation in the form $y = ax + b$, where

b is the intercept on the y-axis and a is the gradient of the line.

You won't find b directly if the line does not extend to the y-axis. Instead, find a first, then calculate b from a particular point, using $b = y - ax$.

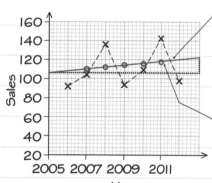

a is the slope. Draw a triangle from two points on the line and then calculate the gradient using

$$\text{gradient} = \frac{\text{height}}{\text{base}}$$
$$= \frac{122 - 106}{8}$$
$$= 2$$

At this point, $y = 116$ and $x = 2011$

So $b = y - ax = 116 - 2 \times 2011$

The equation of the line can then be given as $y = 2x - 3906$

Worked example   tier **H**   Aiming higher

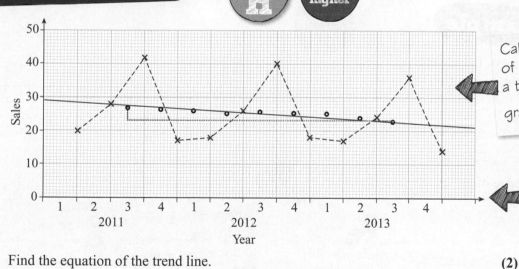

Calculate the gradient of the line by drawing a triangle. Then use $\text{gradient} = \frac{\text{height}}{\text{base}}$

x is measured in **years** not quarters in this example.

Find the equation of the trend line.   (2)

$a = \dfrac{\text{change in } y}{\text{change in } x} = \dfrac{23 - 27}{2 \text{ years}} = -2$

When $x = 2011$, $y = 27$, so $b = y - ax = 27 - (-2) \times 2011 = 4049$

Trend line is $y = -2x + 4049$

Now try this   tier **H**   Aiming higher

1 The diagram shows the number of moths caught in a wood over several years and the 3-point moving averages.
Find the equation of the trend line.   (2)

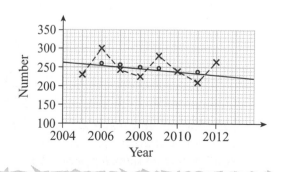

Prediction

You can use the mean seasonal variation and the trend line to make predictions.

Mean seasonal variation

For a time series in which there is a consistent seasonal variation above and below the trend line, you can work out the mean variation for a particular quarter or season.

The mean seasonal variation for the 3rd quarters is $\dfrac{15 + 15 + 14}{3}$

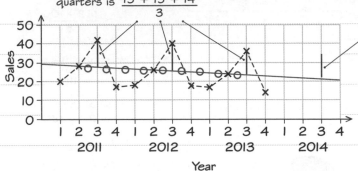

Making predictions

If the seasonal variation behaves in a repeating way, you can predict values beyond the last point in the time series graph.

- Work out the mean variation for that quarter.
- Extend the trend line as far as that quarter.
- Add on or take off the mean variation.

Predicted sales for the 3rd quarter of 2014. In summer the data values are higher than the trend line, so you **add** the mean seasonal variation to predict this.

For the 3rd quarters the variations are positive (above the trend line). They are 15, 15 and 14.

Worked example   tier **H**   Aiming higher

The time series shows the variation in average temperature in France over four years.

(a) Work out the mean seasonal variation for the winter season. **(2)**

Mean seasonal variation for winter

$$= \frac{(-1.5) + (-2.5) + (-2) + (-3)}{4} = -2.25$$

(b) Predict the value of the temperature in winter 2005. **(1)**

$13.8 + (-2.25) = 11.55$

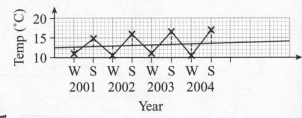

Find the distance between the trend line and the actual temperature for each winter, and add these together. Make sure you include the minus signs because the temperatures are below the trend line so the variation is negative.

Add the seasonal variation you have calculated to the prediction of the temperature from the trend line.

Now try this   tier **H**   Aiming higher

1 The table shows the sales of sun protection lotion sold from a pharmacy for each 4-month period over three years.

Year	Y1			Y2			Y3		
Period	1	2	3	1	2	3	1	2	3
Ice-cream sales	850	1800	950	1000	1900	850	1250	1900	1000
		1200	1250	1283	1250	1333	1333	1383	

Check carefully if the variation is positive or negative so you know whether to add or subtract for the prediction.

(a) Plot the information on a time series graph. **(2)**
(b) Work out the mean variation above the trend line for period 2 in Y1, Y2 and Y3. **(2)**
(c) Predict the sales for period 2 Y4. **(2)**

Probability

Probability is concerned with the chances of outcomes and events.

An **outcome** is the result of a trial (such as rolling a dice once, for which the outcome is the number rolled).

An **event** is one or more outcomes (such as getting a number more than 4 when you roll a dice).

Likelihood

You need to be able to use these words to describe the likelihood of an event happening.

Impossible	Unlikely	Evens	Likely	Certain
(can't happen)		(as likely to happen as not)		(must happen)

Increasing likelihood ⟶

You can also use 'very unlikely' for something that is nearly impossible and 'very likely' for something that is almost certain.

Probability scales

You can also use numbers to represent the probability that an event will occur.

0 is impossible, 0.5 is evens and 1 is certain.

'Likely' is used for events placed here.

Impossible   Evens   Certain

```
|---------------|-------|-------|------|
0               0.5                  1
```

Worked example

tier F

Use a suitable word to describe the likelihood of each of these events.

(a) It will snow in London in May. **(1)**
Unlikely

(b) The next person born in the UK will be a boy. **(1)**
Evens

(c) You roll an ordinary dice and get the number 7. **(1)**
Impossible

◄ Sometimes you will be given a list of words to choose from and then you must choose one from the list.

◄ Numbers on an ordinary dice are 1 to 6.

Now try this

tier F

1 On the probability scale, mark with a cross (×) the probability that
   (a) You will have something to eat tomorrow. Label this cross **A**.
   (b) A teacher chosen at random was born on a Monday. Label this cross **B**.
   (c) A fair 6-sided dice will show an even number when rolled. Label this cross **C**.

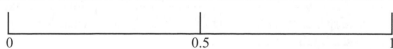

```
|---------------|---------------|
0               0.5             1
```

Sample spaces

A sample space is a list of all the possible outcomes of a trial.

Sample spaces

If a fair dice is rolled once, the sample space is the list of numbers 1, 2, 3, 4, 5, 6.

If the event is 'any even number' then the possible members of the sample space that are included in the event are 2, 4 and 6.

If two fair dice are each rolled once, the sample space consists of all the possible pairs:

(1, 1)   (1, 2)   (1, 3)   (1, 4)   (1, 5)   (1, 6)
(2, 1)   (2, 2)   (2, 3)   (2, 4)   (2, 5)   (2, 6)
(3, 1)   (3, 2)   (3, 3)   (3, 4)   (3, 5)   (3, 6)
(4, 1)   (4, 2)   (4, 3)   (4, 4)   (4, 5)   (4, 6)
(5, 1)   (5, 2)   (5, 3)   (5, 4)   (5, 5)   (5, 6)
(6, 1)   (6, 2)   (6, 3)   (6, 4)   (6, 5)   (6, 6)

If the event is 'any double' then the successful outcomes from the sample space are

(1, 1), (2, 2), (3, 3), (4, 4), (5, 5) and (6, 6).

If the event is 'a total of 4' then the successful outcomes are (1, 3), (2, 2) and (3, 1). You have to include (1, 3) **and** (3, 1) to describe the different orders of results.

Worked example

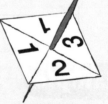

Spinner A   Spinner B

tier F

The diagram shows two spinners.
Spinner A is spun once.
(a) Write down a suitable sample space. **(1)**
1, 1, 2, 3
Spinner A and spinner B are each spun once.
(b) Write down a suitable sample space. **(2)**

		Spinner A			
		1	1	2	3
Spinner B	1	(1, 1)	(1, 1)	(1, 2)	(1, 3)
	1	(1, 1)	(1, 1)	(1, 2)	(1, 3)
	2	(2, 1)	(2, 1)	(2, 2)	(2, 3)
	3	(3, 1)	(3, 1)	(3, 2)	(3, 3)

 Arrange the pairs in a table to make sure you don't miss any successful outcomes.

(c) List all the pairs in the sample space which have a total of 3. **(1)**
There are four pairs: (1, 2), (1, 2), (2, 1) and (2, 1)

There are some repeats in this list. This is not an error: they **should** be included.

Now try this

tier F

1 A 1p coin and a 2p coin are each spun once.
    (a) Write out the sample space. **(1)**
    (b) Which one of 2 heads, 2 tails, or 1 head and 1 tail is the most likely? **(1)**

2 There are three people in a room: Alf, Bert and Carol. Two people are picked from the three.
    (a) Write down all six elements of the sample space. **(1)**
    (b) Which is more likely – two men or a man and a woman? **(1)**

Probability and sample spaces

The probability of an event = $\dfrac{\text{number of successful outcomes}}{\text{total number of possible outcomes}}$

You can use the sample space of all possible outcomes to find the probability of an event.

Simple probability

There are seven beads in the bag.
Three of them are red and four are black.
The sample space shows all the possible
outcomes when a bead is taken out:

B B B B R R R

3 of the 7 outcomes are R, so the probability is $\dfrac{3}{7}$

The probability it will be a red bead is found by
counting the number of outcomes that are red.

EXAM ALERT!

Some candidates will give an answer of $\dfrac{3}{4}$.
Always write the total number of outcomes on the
bottom of the fraction.

> Students have struggled with exam
> questions similar to this – **be prepared!**
>
> Results Plus

More complex sample spaces

When you are looking at the outcome
of two combined trials, it is sometimes
easier to write out the sample space
(or draw a sample space diagram).

☐1☐   ☐2☐   ☐3☐

Two tiles are picked at random from these
three. Once a tile is picked, it can't be
picked again.

The sample space for this event is:

(1, 2) (1, 3)

(2, 1) (2, 3)   ⬅ Each of the answers
is equally likely.

(3, 1) (3, 2)

The probability of picking 1 then 2 is $\dfrac{1}{6}$

The probability of picking 2 then 1 is also $\dfrac{1}{6}$

So the probability of picking 1 and 2 in any

order is $\dfrac{2}{6}$

Worked example

Here are two fair spinners. Each spinner is spun once.
Work out the probability that
(a) spinner A lands on 2 and spinner B lands on 3   **(2)**

Spinner A   **Spinner B**

		Spinner B		
		1	2	3
Spinner A	1	(1, 1)	(1, 2)	(1, 3)
	2	(2, 1)	(2, 2)	(2, 3)
	3	(3, 1)	(3, 2)	(3, 3)

⬅ List all the possible outcomes,
starting with '1', then '2' then
'3'. Arrange them in a table to
make sure you don't miss any.

There are 9 possible outcomes. So the probability is $\dfrac{1}{9}$

(b) spinner A and spinner B land on the same number.   **(1)**

There are 3 successful outcomes out of 9 so the probability is $\dfrac{3}{9} = \dfrac{1}{3}$

⬅ The successful outcomes
are (1, 1), (2, 2) and (3, 3).

Now try this

1 There are 10 students in a room. Seven of them are female. A student is picked at random.
   What is the probability that this student is female?   **(1)**

2 Two fair spinners are each numbered 1 to 4. Each spinner is spun once.
   (a) Write out the sample space.   **(1)**
   (b) Work out the probability that the sum of the numbers the spinners land on is 3.   **(2)**

Venn diagrams and probability

A Venn diagram is a way of showing how items are split between sets.

This Venn diagram shows 35 students, of whom 19 had been shopping at the weekend, 9 had played football at the weekend and 5 had done both.

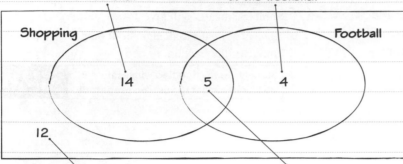

14 people only went shopping at the weekend.

4 people only played football at the weekend.

Shopping   14   5   4   Football

12

$14 + 5 + 4 + 12 = 35$ people in total.

12 people did not play football or go shopping.

5 people went shopping **and** played football.

Worked example

tier **H**   Aiming higher

There are 40 students in a class.
5 study French, German and Spanish.
9 study French and Spanish.
12 study French and German.
7 study German and Spanish.
22 study French.
19 study German.
20 study Spanish.

(a) Draw a Venn diagram. **(3)**

Just French: $22 - (7 + 5 + 4) = 6$
Just Spanish: $20 - 11 = 9$
None: $40 - (9 + 5 + 2 + 7 + 5 + 4 + 6) = 2$

A student is selected at random from the class.
(b) Work out the probability that they study only German. **(1)**

$\frac{5}{40} = \frac{1}{8}$

The diagram shows that 5 students study only German, and there are 40 students in total.

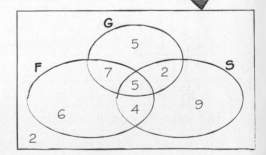

1. Draw the overlapping circles so you can see which sections you need to complete.
2. Write in the numbers you have been given into the correct sections.
3. Calculate the numbers you haven't been given by subtracting them from the ones you know.
4. Don't forget to calculate the number outside the circles, for the students who study no languages.

G
5
F   7   2   S
5
6   4   9
2

Now try this

tier **H**   Aiming higher

1 Adam has a set of 20 tiles: 5 of the tiles have green paint on them, 7 of the tiles have red paint on them, and 3 of the tiles have both red and green paint on them.

(a) Draw a Venn diagram. **(2)**

Adam takes one of the tiles at random.

(b) Work out the probability that the tile has no red paint and no green paint on it. **(2)**

Mutually exclusive events

Mutually exclusive events cannot **both** occur at the **same time**.

Writing probabilities

The probability of rolling a 6 is $\frac{1}{6}$
You can write $P(6) = \frac{1}{6}$
There is one 6. There are six possible outcomes: 1, 2, 3, 4, 5, 6.

The probability of a coin landing heads up is $\frac{1}{2}$. You can write $P(\text{head}) = \frac{1}{2}$
There is one head. There are two possible outcomes: heads or tails.

The sum of probabilities

The probabilities of all the possible outcomes of an event add up to 1.

If you know the probability that something will happen, you can calculate the probability that it won't happen.

P(event A doesn't happen)
= 1 − P(event A happens)

This is written as: $P(\text{not } A) = 1 - P(A)$

Formula for mutually exclusive events

If two events A and B are mutually exclusive then $P(A \text{ or } B) = P(A) + P(B)$.

This spinner is spun once.

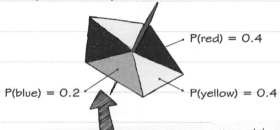

P(red) = 0.4
P(blue) = 0.2
P(yellow) = 0.4

The three outcomes (red, yellow, blue) are **exhaustive** (no others are possible) so their probabilities add up to 1.

The formula gives further information on the probabilities:

P(red or yellow) = P(red) + P(yellow)
= 0.4 + 0.4 = 0.8
P(red or blue) = P(red) + P(blue)
= 0.4 + 0.2 = 0.6
P(blue or yellow) = P(blue) + P(yellow)
= 0.2 + 0.4 = 0.6

Worked example

tier **F&H**

The table gives some information about the probability of taking a bead of a given colour from a box.

Colour	Red	Green	Blue	Yellow
Probability	0.1	0.25	x	0.3

(a) Work out the value of x.   **(2)**

$x = 1 - (0.1 + 0.25 + 0.3) = 0.35$

(b) Work out the probability that a green bead or a yellow bead is taken.   **(1)**

$0.25 + 0.3 = 0.55$

The bead cannot be green **and** yellow so they are mutually exclusive events. So you use P(green or yellow) = P(green) + P(yellow).

Now try this

tier **F&H**

1 Susie has red, black, blue and pink jumpers in her wardrobe. One day she selects a jumper at random. The table gives information about the probability that she selects a jumper of a given colour.

Colour	Red	Black	Blue	Pink
Probability	0.15	0.25	0.32	x

(a) Work out the value of x.   **(2)**
(b) Work out the probability that she selects a red jumper or a black jumper.   **(1)**

Independent events

If the outcome of one event does not affect the probability of another event occurring, the two events are independent.

Formula for independent events

If two events A and B are independent, then P(A and B) = P(A) × P(B).

 A dice is rolled twice.

The probability of rolling a 4 and then a 6 is:

$$P(4 \text{ then } 6) = P(4) \times P(6) = \frac{1}{6} \times \frac{1}{6} = \frac{1}{36}$$

 A coin is spun twice.

The probability of getting two heads

$$P(H \text{ and } H) = P(H) \times P(H) = 0.5 \times 0.5 = 0.$$

> You revised how you could also work out probabilities like these by drawing a **sample space diagram** on page 68.

> The outcome of the first spin does not affect the outcome of the second spin, so the events are independent.

Worked example

tier F&H

> The events 'getting a 4' and 'getting a head' are independent. So you use
> P(4 and head) = P(4) × P(head).

Clemmie throws a fair dice and a fair coin.
(a) Work out the probability that Clemmie gets a 4 and a head. **(2)**

$$\frac{1}{6} \times \frac{1}{2} = \frac{1}{12}$$

(b) Work out the probability that Clemmie gets an even number and a head. **(2)**

$$\frac{3}{6} \times \frac{1}{2} = \frac{3}{12} = \frac{1}{4}$$

Now try this

tier F&H

1 Tom has a spinner with several different coloured sectors of different sizes.
The table gives some information about the probability of getting each colour on the spinner when it is spun once.

Colour	Purple	Blue	Green	Pink
Probability	0.05	x	0.25	0.4

Tom spins the spinner once.
(a) Work out the probability that the spinner lands on blue. **(1)**

Tom spins the spinner twice.
(b) Work out the probability that the spinner stops on pink then green. **(2)**
(c) Work out the probability that the spinner stops on purple both times. **(2)**

Probabilities from tables

You need to be able to calculate probabilities from data given in two-way tables.

The table shows the sandwich orders at a deli one lunchtime. Each person chose one sandwich.

		Bread		
		White	Brown	Total
Filling	Turkey	4	5	9
	Jam	6	4	10
	Cheese	7	6	13
	Total	17	15	32

10 people chose jam.
So if a person is picked at random, the probability that they chose jam is

$$P(jam) = \frac{\text{number of successful outcomes}}{\text{total number of possible outcomes}}$$

$$= \frac{10}{32}$$

$$= 0.3125$$

7 people chose cheese on white bread.

This is the total number of people. So if one person is picked at random, this is the total number of possible outcomes.

Worked example

tier F

This table gives information about the services done by a garage at the start of the year.

		Month			
		Jan	Feb	Mar	Total
Service	Tyres	15	8	4	27
	Exhaust	19	22	7	48
	Total	34	30	11	75

One of these services is selected at random for a quality check.

(a) Which service, tyres or exhaust, is more likely to be selected? **(1)**

Exhaust

(b) Write down the probability that the service was done in February. **(1)**

$\frac{30}{75}$

(c) Write down the probability that the service was tyres and it was done in January. **(1)**

$\frac{15}{75} = \frac{1}{5}$

The most likely service is the one with the higher total. There were 48 exhaust services and 27 tyre services. So an exhaust service is more likely to be selected.

To find the probability that the service was done in February use:

$$P(Feb) = \frac{\text{number of successful outcomes}}{\text{total number of possible outcomes}}$$

$$= \frac{30}{75}$$

If you like, you can simplify the fraction:

$$= \frac{2}{5}$$

There were 15 tyre services in January (number of successful outcomes) and 75 services done in total (total number of possible outcomes). So the probability is $\frac{15}{75}$.

Now try this

tier F

1 This table gives information about some students' eating habits.

		Year group			
		Year 9	Year 10	Year 11	Total
Diet	Vegetarian	15	8	4	
	Not vegetarian	19	22	7	48
	Total			11	75

Look back at page 14 to remind yourself how to complete two-way tables.

(a) Complete the table. **(2)**

One student is selected at random for a further interview.

(b) (i) Write down the diet that this student is most likely to have. **(1)**

(ii) Write down the probability that a student from year 10 will be selected. **(1)**

(iii) Write down the probability that a vegetarian student from year 11 will be selected. **(1)**

Experimental probability

You can find estimates of a probability by repeating an experiment many times.

The formula for the experimental probability of an event is:

number of times the event happens
  total number of trials

The more times the experiment is repeated, the more accurate the estimate can be.

The table shows the results of an experiment to discover whether a coin is biased.

Number of heads	34
Number of throws	50
Estimate	0.68

The results give an experimental probability of $\frac{34}{50} = 0.68$ for getting a head.

If you repeated this test, you could get a more accurate result for the experimental probability by dividing the total number of heads by the total number of throws.

Worked example

tier F&H

Sharon has a biased dice. She knows that the probability of getting a 6 is $\frac{1}{5}$ rather than $\frac{1}{6}$. She throws the dice 180 times. Work out an estimate for the number of 6s that she can expect. **(1)**

Estimate = 180 × 0.2 = 36

Golden rule

Estimate = total number of trials × probab

Experimental vs predicted

You can compare predicted results with results from an experiment to test for **bias**.

This table shows the predicted and experimental results of spinning a 4-sided spinner 100 times.

	1	2	3	4
Predicted	25	25	25	25
Experimental	21	22	33	24

The experimental figures are higher than expected for the number 3. The spinner could be biased towards 3.

Now try this

tier F&H

1 Suha spins a coin 250 times. She gets 110 heads.
   (a) Work out the experimental probability of getting a head. **(2)**
   (b) Write down the experimental probability of getting a tail. **(1)**

2 Abel, Beth and Connor each test the same dice for bias. Their results are given in the table.

	Abel	Beth	Connor
Number of 2s	15	24	36
Number of rolls	60	90	150
Estimate of probability of a 2			

Remember the mor data, th more accurac

Connor claims that his results will give the most accurate estimate.
   (a) Explain why Connor is correct. **(1)**
   (b) Work out what Connor's estimate should be. **(2)**
   (c) (i) Explain how it is possible from these results to get a better estimate than Connor's.
       (ii) Find the value of the best estimate. **(2)**

Risk

Risk and accidents

The risk of a particular type of accident or problem with a machine is obtained from data. The risk of an accident is its experimental probability.

Write the risks as decimals. It makes them easier to compare.

This table shows the number of times some basketball players acquired injuries when playing.

Injury	Fingers	Feet	Knees
Number of injuries	6	8	3
Games played	50	46	50

The risk of an injury to fingers is $\frac{6}{50} = 0.12$

The other two injury types have risks of 0.17 and 0.06

Cost of risk

Insurance companies try to evaluate the cost of events happening. This is known as the **cost of risk**. Insurance companies use it to work out what premiums to charge.

From previous cases, an insurance company knows that the risk (experimental probability) of a freezer breaking down in a given year is 0.02. They also know that the typical cost when a freezer breaks down is £200.

Cost of the risk = £200 × 0.02 = £4 per year

Worked example
(tier F&H)

The table gives information about the number of faults reported in some electrical goods.

	Iron	Kettle	Microwave
Faults	28	15	23
Number sold	2000	2500	1560

Work out the risk of a fault for each electrical good. **(3)**

Iron $\frac{28}{2000} = 0.014$, kettle $\frac{15}{2500} = 0.006$,

microwave $\frac{23}{1560} = 0.015$

Golden rule

If you know the risk of a fault then you can predict the number of future faults:

number of faults

= risk of a fault × number of items sold

EXAM ALERT!

Many students do not understand the idea of risk. Remember that it is just the same as experimental probability.

Students have struggled with exam questions similar to this – **be prepared!**

 Results Plus

Now try this
(tier F&H)

1 The table shows the number of knee injuries in some games in three sports in the 2014 season.

	Football	Hockey	Rugby
Injuries	8	5	13
Games	50	60	40

(a) Work out the risk of a knee injury in each sport. **(3)**

Leave your answers as decimals.

(b) 35 hockey games are going to be played in the 2015 season.
   (i) Estimate the number of knee injuries next season. **(1)**

You need to predict the number of future injuries.

An insurance company estimates that the average cost of operating on a knee injury is £400.

(c) Calculate how much premium they need to charge per hockey game to cover their likely costs. **(2)**

Cost of the risk = cost of fix × experimental probability of the accident

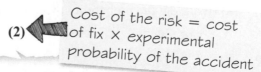

75

Probability trees

Independent events

The tree diagram shows all the possible outcomes when one counter is picked from bag A and one counter is picked from bag B. The outcome of one pick does not affect the outcome of the other pick so the two events are **independent**. There is more about independent events on page 72.

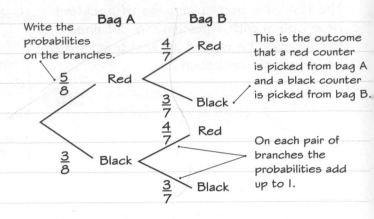

Write the probabilities on the branches.

$\frac{5}{8}$ Red

$\frac{4}{7}$ Red — This is the outcome that a red counter is picked from bag A and a black counter is picked from bag B.

$\frac{3}{7}$ Black

$\frac{3}{8}$ Black

$\frac{4}{7}$ Red

$\frac{3}{7}$ Black — On each pair of branches the probabilities add up to 1.

Worked example

tier F&H

Bag X contains 5 black beads and 3 white beads.
Bag Y contains 2 black beads and 3 white beads.
Simon takes a bead at random from each bag.
The probability tree diagram gives some information about the probabilities.

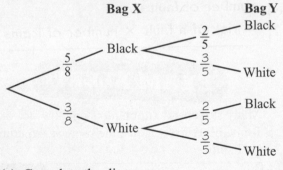

Bag X   **Bag Y**

$\frac{5}{8}$ Black $\frac{2}{5}$ Black $\frac{3}{5}$ White

$\frac{3}{8}$ White $\frac{2}{5}$ Black $\frac{3}{5}$ White

(a) Complete the diagram. **(2)**

(b) Work out the probability that both beads are black. **(2)**

$$\frac{5}{8} \times \frac{2}{5} = \frac{10}{40}$$

(c) Work out the probability that both beads are the same colour. **(2)**

$$\frac{5}{8} \times \frac{2}{5} + \frac{3}{8} \times \frac{3}{5} = \frac{19}{40}$$

Golden rule

To work out probabilities in a tree diagram you:

MULTIPLY ALONG THE BRANCHES

ADD UP THE OUTCOMES

You **multiply** along the branches on a tree diagram to work out the probability that **both** events occur.

There are two possible outcomes: black, black and white, white. When there is more than one possible outcome from a tree diagram, you **add** the relevant branches to work out the probability that **one** of the events might occur.

Now try this

tier F&I

1 Bag X contains 4 black beads and 3 red beads. Bag Y contains 2 black beads and 1 red bead.
   Susan takes a bead at random from each bag.
   (a) Draw a suitable probability tree diagram. **(2)**
   (b) Work out the probability that one bead is black and one bead is red. **(2)**

Conditional probability

The probability of event A occurring **given that** event B has already occurred is denoted by P(A|B).

Look out for the key word 'given' – this implies **conditional probability.**

The probability of events A and B **both** occurring is P(A|B) × P(B).

Events X and Y are **not** independent, as the outcome of the first pick affects the probabilities for the second pick. This is known as **sampling** (or selection) **without replacement** and always involves conditional probabilities.

Use the multiplication rule to find probabilities in tree diagrams.

Worked example

tier **H**

A bag contains 4 black beads and 3 red beads. One bead is picked at random, not replaced, and then a second bead is picked at random.
X is the event 'the first bead is black'.
Y is the event 'the second bead is black'.

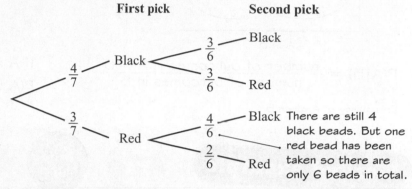

There are still 4 black beads. But one red bead has been taken so there are only 6 beads in total.

(a) Find the probability that the second bead is black, given that the first bead is black.

$$P(Y|X) = \frac{3}{6}$$

(b) Find the probability that both beads are black.

$$P(Y|X) \times P(X) = \frac{3}{6} \times \frac{4}{7} = \frac{12}{42}$$

Worked example

tier **H** **Aiming higher**

The probability of a person having a peanut allergy is 0.1.
A test is available to see if a person has this allergy.
The result for a person with the allergy is 80% accurate.
The result for a person without the allergy is 95% accurate.
Complete the tree diagram for the two events.   **(3)**

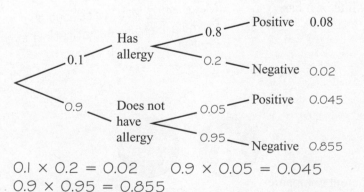

$0.1 \times 0.2 = 0.02$ $0.9 \times 0.05 = 0.045$
$0.9 \times 0.95 = 0.855$

EXAM ALERT!

Read the information in the question carefully to work out the probabilities on each set of branches – don't assume they are the same.

Students have struggled with exam questions similar to this – **be prepared!**   **Plus**

With conditional probability the second pairs of branches will have different probabilities on them as they each depend on the first outcome.

You can calculate the missing numbers because each set of branches must add up to 1.

Now try this

tier **H** **Aiming higher**

1 There are 6 black beads and 4 red beads in a bag. Two beads are taken at the same time.
   Work out the probability that both beads will be the same colour.   **(3)**

Probability formulae

You need to learn and remember these formulae and notations:

$P(A \cap B) = P(A|B) \times P(B)$

This means P(A **and** B).

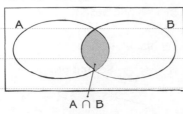

A ∩ B

$$P(A|B) = \frac{\text{number of outcomes in } A \cap B}{\text{number of outcomes in } B}$$

Probability of A **given** B.

$P(A \cup B) = P(A) + P(B) - P(A \cap B)$

This means P(A **or** B).

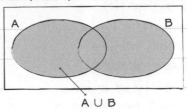

A ∪ B

If A and B are mutually exclusive, then:

$P(A \cup B) = P(A) + P(B)$

Worked example

tier **H**   **Aiming higher**

Kim has a set of 20 tiles. 10 of the tiles have the letter X on them, of which 6 are red. The other tiles have the letter Y on them, of which 7 are red. All other tiles are white.
Kim takes a single tile at random.
Let A be the event 'a white tile is selected'.
Let B be the event 'a tile with the letter X is selected'.

(a) Find

   (i) P(A)

$(10 - 6) + (10 - 7)$

$= 4 + 3 = 7$ white tiles out of a total of 20 tiles

$P(A) = \frac{7}{20}$

   (ii) P(B|A)

$P(B|A) = \frac{\text{number of outcomes in } A \cap B}{\text{number of outcomes in } A} = \frac{4}{7}$

   (iii) P(A ∩ B)   **(3)**

$P(A \cap B) = P(B|A) \times P(A) = \frac{4}{7} \times \frac{7}{20} = \frac{4}{20}$

(b) Work out the probability that the tile is white or it has the letter X on it, or both.   **(2)**

$P(A \cup B) = P(A) + P(B) - P(A \cap B)$

$\qquad = \frac{7}{20} + \frac{10}{20} - \frac{4}{20} = \frac{13}{20}$

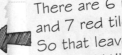
There are 6 red tiles out of 10 Xs and 7 red tiles out of 10 Ys. So that leaves 4 white X tiles and 3 white Y tiles.

If you have to calculate a conditional probability like P(B|A) then you always need to use the formula for conditional probability. Make sure you write out the formula before substituting the probabilities you know.

Now try this

tier **H**   **Aiming higher**

1 JLD Engineering gets parts from two different companies.
  70% of these parts are supplied by Company A and 30% are supplied by Company C.
  The quality of the parts from Company A is 97% good and 3% bad.
  The quality of the parts from Company C is 95% good and 5% bad.
  Let A denote the event 'a part from company A'. Let G denote the event 'good quality'.
  (a) Draw a probability tree diagram.   **(3)**
  (b) Find (i) P(A), (ii) P(G|A), (iii) P(A ∪ G), (iv) P(G).   **(4)**

Simulation

You can simulate situations involving probabilities by using random numbers.

If you know these probabilities of different types of vehicle arriving at a road junction, the second table shows how you might model the traffic.

You can use a random number generator (on a calculator or a computer) to produce a sequence of numbers that simulates the situation.

Vehicle type	Van	Car	Lorry	Other
Probability of arrival	0.2	0.5	0.2	0.1

Use the ten numbers from 0 to 9. Assign them to the vehicle types, according to the probability they will arrive at the junction.

Vehicle type	Van	Car	Lorry	Other
Random numbers	0, 1	2–6	7, 8	9

The one number 9 has been assigned to 'Other' because the probability is $0.1 = \frac{1}{10}$

The two numbers 0 and 1 have been assigned to 'Van' because the probability is $0.2 = \frac{2}{10}$

The sequence of random numbers 8, 8, 4, 5, 0, 2, 0, 3 would simulate Lorry, Lorry, Car, Car, Van, Car, Van, Car.

Worked example

The table gives information about the proportions of types of people arriving at a bus stop.

Type	Proportion
Man	35%
Woman	40%
Boy	17%
Girl	8%

Interpret this as 35 out of every 100, which is a probability of $\frac{35}{100}$

(a) Assign suitable random numbers to simulate arrivals. **(1)**

Type	Man	Woman	Boy	Girl
Numbers	00–34	35–74	75–91	92–99

(b) Use the first 12 of the 2-digit random numbers below to simulate the first 12 arrivals.

95   50   43   03   06   69   71   52   80

21   41   32   06   70   93   95   45   94   **(1)**

Girl, Woman, Woman, Man, Man, Woman, Woman, Woman, Boy, Man, Woman, Man

EXAM ALERT!

Many candidates get confused and give 0–39 for Woman (trying to show 40%). It should be 35–74.

Students have struggled with exam questions similar to this – **be prepared!** Results**Plus**

 The number 00 is included, so 'man' goes up to 34, **not** 35.

The first arrival is a girl, as 95 falls in the range 92–99.

Now try this

1 The table gives the probabilities of 0, 1 or 2 people joining a queue in any minute interval.

Number joining	0	1	2
Probability	0.3	0.6	0.1

(a) Explain how to use random digits from 0 to 9 to simulate the arrivals. **(1)**

(b) Use these random digits to simulate the first 10 minutes of a queue. **(1)**

0   1   0   1   8   9
6   8   8   8   9   4
9   2   0   4   2   6

(c) If nobody leaves the queue, use your answer to part (b) to estimate its length after 10 minutes. **(1)**

Probability distributions

The probability distribution of X is the set of values X can take and the associated probabilitie

Probability distributions are a way of showing all the outcomes and their probabilities. The outcomes have to be **numerical**, such as the number of heads when three fair coins are spun.

X is the number of heads

X	0	1	2	3
P(X)	0.125	0.375	0.375	0.125

The probabil MUST add u to 1

The probability of getting exactly 1 head is 0.375

Worked example

tier **H**   Aiming higher

Probabilities add up to 1.

Kevin rolls a biased dice once.
The table gives the probability distribution of X, the number the dice lands on.

X	1	2	3	4	5	6
P(X)	0.1	k	k	k	k	0.1

(a) Work out the value of k.   (2)

$0.1 + k + k + k + k + 0.1 = 1$
$k = 0.2$

(b) Work out the probability that $X > 4$.   (2)

$0.2 + 0.1 = 0.3$

Even if you get k wrong you could still get credit for a correct method, so always show your working.

$X > 4$ is the same as $X = 5$ or $X = 6$ so find the probability that X is one o these values.

The discrete uniform distribution

In a discrete uniform distribution all the values of X have the same probability. If there are N possible outcomes then the probabilities are all $\frac{1}{N}$

The score on a **fair** dice has a uniform distribution with all probabilities $\frac{1}{6}$

The probability distribution of X, the number that this spinner lands on, is uniform discrete with all probabilities 0.25

X	P(X)
1	0.25
2	0.25
3	0.25
4	0.25

Now try this

tier **H**   Aiming higher

1 The table gives the probability distribution of the total score, X, when two fair four-sided spinners, labelled 1, 2, 3 and 4, are spun.

X	2	3	4	5	6	7	8
P(X)	$\frac{1}{16}$	$\frac{2}{16}$	k	$\frac{4}{16}$	k	$\frac{2}{16}$	$\frac{1}{16}$

(a) Work out the value of k.   (2)
(b) Find the probability that $X < 6$.   (2)

The binomial distribution 1

Probability and repeated trials

Two biased **identical** coins are each spun once.

The possible outcomes are: 2 heads (HH)
$\qquad\qquad$ 1 head and 1 tail (HT)
$\qquad\qquad$ 2 tails (TT)

The probabilities are: $P(HH) = P(H) \times P(H)$
$\qquad\qquad P(HT) = 2 \times P(H) \times P(T)$
$\qquad\qquad P(TT) = P(T) \times P(T)$

where P(H) is the probability of heads on one coin and P(T) is the probability of tails on one coin, no matter what the bias is.

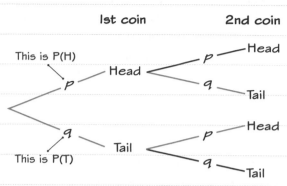

The blue lines in the probability tree diagram show the two ways of getting 1 head and 1 tail, with total probability $p \times q + q \times p = 2pq$

Probability and binomials

You can use the expansion of

$(p + q)^2 = p^2 + 2pq + q^2$

to work out probabilities when there are two repeated trials of two events that are mutually exclusive.

> In place of heads, think of **successes** ☺.
> In place of tails, think of **failures** ☹.
> In place of throws, think of **trials**.
> The probability of a success in any trial is p.
> The probability of a failure in any trial is q.
> Since success or failure are the only two outcomes in a single trial, $p + q = 1$.

Worked example

tier H

Two bags, A and B, each contain 5 white and 3 red beads. A bead is taken at random from bag A and a bead is taken at random from bag B.

(a) Find the probability that both beads are white. **(2)**

Let getting white be a success.

$p = \dfrac{5}{8}$ and $q = \dfrac{3}{8}$

Probability of two white beads $= p^2 = \dfrac{25}{64}$

(b) Find the probability that exactly one bead is white. **(2)**

Probability $= 2 \times \dfrac{5}{8} \times \dfrac{3}{8} = \dfrac{30}{64}$

◀ The conditions for the two bags are **identical**.

◀ Find q from $q = 1 - p$.

◀ The probability of two successes out of two is p^2.

◀ The probability of one success and one failure out of two trials is $2pq$.

Now try this

tier H · **Aiming higher**

1. Alice has a biased dice. The probability of getting 6 on any roll is 0.2. Alice rolls the dice twice.
   (a) Work out the probability that she gets exactly one 6. **(2)**
   (b) Work out the probability that she does not get a 6 in either of the rolls. **(2)**

The binomial distribution 2

When the number of trials is more than two it becomes more difficult to draw probability tree diagrams.

Instead, you can use the **binomial expansion** to find the probability distribution.

For three, four and five trials, the expansions are:

3 trials
$$(p + q)^3 = p^3 + 3p^2q + 3pq^2 + q^3$$

4 trials
$$(p + q)^4 = p^4 + 4p^3q + 6p^2q^2 + 4pq^3 + q^4$$

5 trials
$$(p + q)^5 = p^5 + 5p^4q + 10p^3q^2 + 10p^2q^3 + 5pq^4 + q$$

You do not have to remember any of these formulae.
The one you need to use will be given in the question.

Using the expansions

For **three trials** the probability distribution of X, the number of successes, is:

X	3	2	1	0
$P(X)$	p^3	$3p^2q$	$3pq^2$	q^3

where p is the probability of success and q is the probability of failure on any one trial, and $p + q = 1$

If the probability of success is $p = 0.7$ then the formula can be used to work out the probability of one success and two failures in three trials.

Since $p + q = 1$, $q = 1 - 0.7 = 0.3$

$P(X = 1) = 3pq^2 = 3 \times 0.7 \times 0.3^2 = 0.189$ ◀

Repeated trials

You can use a binomial distribution whe you have a sequence of repeated trial in which:

- there are only two outcomes – know as success and failure
- the probability of a success in each trial is constant, traditionally denoted by p
- each trial is independent of all other trials.

To decide which term in the expansion to use, match the value of X to the power of p.

Worked example

tier H

In the question you may use
$(p + q)^4 = p^4 + 4p^3q + 6p^2q^2 + 4pq^3 + q^4$
On a spinner, the probability of landing on blue is 0.4.
The spinner is spun four times.
Let X be the number of times the spinner lands on blue.
(a) Work out $P(X = 2)$.   **(2)**
$p = 0.4$, so $q = 0.6$
$P(X = 2) = 6 \times 0.4^2 \times 0.6^2 = 0.3456$ ◀
(b) Work out $P(X > 2)$.   **(2)**
$P(X > 2) = 4 \times 0.4^3 \times 0.6 + 0.4^4$ ◀
$= 0.1536 + 0.0256 = 0.1792$

The correct formula will always be given. You have to know how to **use** the formula by selecting the correct value of p and then the correct term.

Use the term which has the same power of p as the number of successes.

$X > 2$ means that $X = 3$ or $X = 4$. These two cases are mutually exclusive so the probabilities can be added.

Now try this

tier H   **Aiming higher**

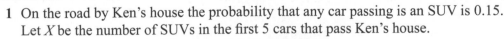

1   On the road by Ken's house the probability that any car passing is an SUV is 0.15.
    Let X be the number of SUVs in the first 5 cars that pass Ken's house.
    (a)   Work out the probability that all 5 cars are SUVs.   **(1)**
    (b)   Work out the probability that at least 4 cars are SUVs.   **(2)**
    Use $(p + q)^5 = p^5 + 5p^4q + 10p^3q^2 + 10p^2q^3 + 5pq^4 + q^5$

The normal distribution

X has a normal distribution if its graph looks like this.

The graph is symmetrical about the mean value of X, denoted by μ.

The standard deviation of X is denoted by σ.

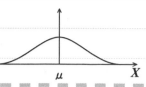

The normal distribution

Different values of μ and σ give different normal distributions but they all have the following properties:

- Always symmetrical about the mean μ.

- The probability that X lies within $\pm 2\sigma$ of the mean is 0.95. So 95% of all observations of X should give a value between $\mu + 2\sigma$ and $\mu - 2\sigma$.

- The probability that X lies within $\pm 3\sigma$ of the mean is 0.998, so almost certain. Virtually all observations of X (99.8%) should give a value between $\mu + 3\sigma$ and $\mu - 3\sigma$.

The normal distribution can be used to model many situations, including:

- people's heights
- people's blood pressure
- people's body mass index
- size of car parts from a factory.

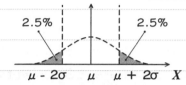

2.5%   2.5%

$\mu - 2\sigma$ μ $\mu + 2\sigma$ X

EXAM ALERT!

Many students forget these values. They are not on the formulae sheet so you must learn them.

Students have struggled with exam questions similar to this – **be prepared!**   Results Plus

Worked example

 tier **H** Aiming higher

The heights of a species of daffodil are normally distributed.

2.5% of the heights are greater than 16.5 cm.

50% of the heights are greater than 13.5 cm.

Find the mean and the standard deviation.   **(2)**

$\mu = 13.5$

$\mu + 2\sigma = 16.5$ so $\sigma = (16.5 - 13.5) \div 2 = 1.5$

The distribution is symmetrical about the mean so 50% must be above the mean.

95% lie within $\mu + 2\sigma$ and $\mu - 2\sigma$ so 2.5% are more than $\mu + 2\sigma$.

Worked example

tier **H** Aiming higher

A doctor records the BMI of her 50-year-old male patients.

The BMI is normally distributed with a mean of 28 and a standard deviation of 4.

Calculate the lowest BMI number for her patients.

Justify your answer.   **(2)**

$\mu - 3\sigma = 28 - 3 \times 4 = 16$

99.8% of the population lie within 3 standard deviations of the mean so the lowest BMI will be very close to the mean − 3 standard deviations.

Now try this

tier **H**

1 The masses of adult trout are normally distributed with mean 2.4kg and standard deviation 0.2kg.
   (a) Work out the mass which is exceeded by 2.5% of this trout population.   **(2)**
   (b) Between what values would almost all of the trout masses lie?   **(2)**

Quality control 1

Companies have to take steps to ensure that the quality of their products is satisfactory. They can do this by taking samples of their product periodically and then using suitable statistical techniques on each sample.

Control charts and sample means

When items are made on a production line, the company needs to check their quality. One way of doing this is to take a sample periodically and test the items in the sample – by weighing or by measuring lengths, for example.

A tester will work out the **sample mean** of the measurements and plot it on a **quality control chart**.

Each sample will have a different mean but the distribution of the sample means is generally normal with mean μ.

The warning limits are $\pm 2\sigma$ from the target mean μ.

The action limits are $\pm 3\sigma$ from the target mean μ.

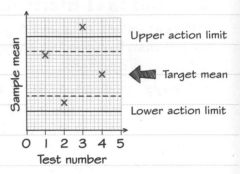

Actions to ensure quality control

1 The sample mean is within the warning limits – do nothing.

2 The sample mean is between the warning limit and the action limit – do a second test immediately.

3 The sample mean is outside the action limit. Stop the machine, check it and reset.

If everything is working correctly:
- values of the sample mean should lie within $\pm 2\sigma$ of the target mean μ, in 19 out of 20 samples (95%)
- values of the sample mean should lie within $\pm 3\sigma$ of the target mean in 998 out of 1000 samples (99.8%), so in this case the probability of being outside these limits is 0.2% (assuming the machine is working properly).

Worked example

tier H **Aiming higher**

In order to check the lengths L mm of precision needles, a factory tests samples of 10 needles every 30 minutes. For the first three samples, the values were
1. $\sum L = 180$ 2. $\sum L = 183.5$ 3. $\sum L = 185$
(a) Plot the means for each of the samples on a control chart. **(2)**
1. Mean = 18 2. Mean = 18.35 3. Mean = 18.5
(b) State what action, if any, should be taken. **(2)**
1. No action
2. Test again
3. Stop the process and reset. ⬅ Always make it clear which sample you are discussing. For sample 2 the mean lies between the warning and the action limits.

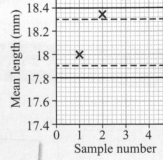

Now try this

tier H **Aiming higher**

1 A machine fills bottles with liquid. 12 bottles with volume V ml of liquid in them are tested every 10 minutes. The distribution of the sample means is normal with $\mu = 330$ ml and $\sigma = 2.4$ ml.
  (a) Work out warning and action limits for this sampling process. **(2)**
  (b) One sample has $\sum V = 3398$. What action should be taken? **(2)**

Quality control 2

Quality control charts can also use the median or the range when monitoring the quality of goods.

In each case the process is similar – samples are taken periodically and tested. Action, if any, is then taken on the basis of the test results.

Control charts and sample medians

The control chart for medians looks the same as that for means.

For each test the value of the median of the items in the sample, the **sample median**, is plotted.

Warning limits are $\pm 2\sigma$ from the target median.

Action limits are $\pm 3\sigma$ from the target median.

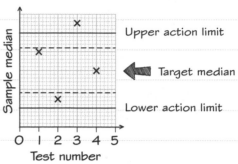

Control charts and sample ranges

As well as monitoring the mean or median value of samples from a production line, companies also monitor the variability of the items on the line.

It is possible, for example, when producing biscuits that the average weight is on target but the **variability** of the weights is unacceptable.

The **range** is used to measure the variability, as it is much quicker to find than the standard deviation.

The control chart for ranges looks the same and has the same limits as for medians and means.

A lower limit is sometimes shown, as **no** variation may mean that the testing equipment is not working.

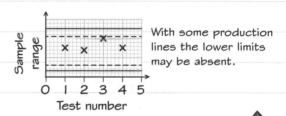

With some production lines the lower limits may be absent.

There is no simple relationship between the sample range and the normal distribution. The warning and action limits are based on the testing itself.

Worked example

The diagram shows a control chart for sample ranges in a biscuit factory. Some tests have already been done.

(a) Should the line have been stopped at any time? **(1)**

No, because all the ranges are within the warning limits.

For sample number 4, the weights in grams were

15.2   14.9   15.3   15.3   15.0   15.1   14.9   15.1

(b) Complete the control chart for sample 4. **(2)**

$15.3 - 14.9 = 0.4$

(c) State what action, if any, should be taken after sample 4. **(2)**

No action needed, as still within the warning limits.

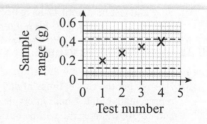

Although it is easy to do in your head, you should always show the calculation of the range.

Now try this

1   A control chart for medians is used to monitor a production line that makes cakes.
The target weight is 125 g and $\sigma = 3.5$ g.
Work out the upper and lower warning and action limits for the control chart. **(3)**

Answers

COLLECTING DATA

1. Types of data
(a) Sets B and C   (b) Sets A and D
(c) Set C   (d) Set B
(e) Sets A and D

2. Measurements and variables
1. (a) 23.5 m²   (b) 24.5 m²
2. (a) Both quantitative, age
   (b) Both quantitative, age

3. Sampling frames, pre-tests and pilots
1. (a) It is biased as his friends are likely to do similar things.
     Its size is too small.
   (b) He should take a random sample of 30 or more using a school list of all the students in his year.

4. Experiments and hypotheses
1. Select two sets of plants of the same type at random.
Put both sets in the same conditions.
Spray only one set with detergent.
Count the number of insects on each plant of both sets at the same time.
2. 1138
3. 76%

5. Stratified sampling
1. 27

6. Further stratified sampling
1. 14

7. Further sampling methods
1. (a) Select a number n from 1 to 4 at random. Then select the nth house, the $(n + 4)$th house, the $(n + 8)$th house …
   (b) This takes one house from each block and in the same position in each block so is not fully random.
   (c) Number all the houses in the street and select by using a list of random numbers.
2. (a) Because it is quick and cheap to do.
   (b) Stop and test men until 12 have been interviewed. Stop and test women until 28 have been interviewed.

8. Sampling overview
1. (a) Systematic
   (b) He may get the cars at times of day which are not representative: the 4-hour intervals could be at 8:30, 12:30 and 16:30, when traffic will be busier.
2. (a) If there are different groups in the population which could be very different in character and have different sizes.
   (b) It should take a stratified sample in which the number of older voters sampled is 3 times the number of younger voters sampled.

9. Data capture sheets
1.

Type of person	Tally	Frequency
Man		
Woman		
Boy		
Girl		

2.

Type of precious stone	Tally	Frequency
Ruby	卌 |||	8
Diamond	卌	5
Emerald	卌|	6
Sapphire	卌	5

10. Interviews and questionnaires

Interview   Advantage – allows the doctor to explain questions
              Disadvantage – respondents may not be completely honest face to face with the doctor

Questionnaire   Advantage – can be done anonymously
              Disadvantage – some respondents may not fully understand the questions

11. Questionnaires
(a) The question is too open.
The response boxes allow only favourable responses.
(b) An improved question would be:
What did you think of the quality of the service you received today?

Very poor ☐
Poor ☐
Neither poor nor good ☐
Good ☐
Very good ☐

12. Capture/recapture
1. (a) 520
   (b) One of: that the marked bears are no more likely to be captured than the unmarked bears; that the marked bears are no more likely to die than the unmarked bears; that the population of marked bears does not change between the two samples.

REPRESENTING DATA

13. Frequency tables
1.

Time, (T seconds)	Tally	Frequency
$1 < T \leq 10$	|||	3
$10 < T \leq 20$	卌|	6
$20 < T \leq 30$	卌|||	8
$30 < T \leq 40$	|||	3

14. Two-way tables
1. (a)

	School	Packed	Home	Total
Boys	10	4	3	17
Girls	7	4	2	13
Total	17	8	5	30

(b) 8

15. Pictograms
1.

Monday	▦
Tuesday	▦ ▦ ▦
Wednesday	▦ ▦
Thursday	▦ ▦ ▦

Key

 represents 4 boxes

16. Bar charts and vertical line graphs

1.

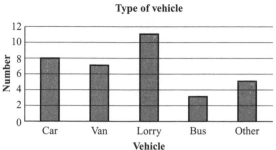

17. Stem and leaf diagrams

1.

10	4 5 8
11	0 2 3 9
12	0 1 2 2 7
13	1 2 3 7

Key 10|4 represents 104 cm

2.

3	8 9
4	1 7 7 9
5	3 5 6
6	0 1 3 4
7	1 2

Key 3|8 represents 3.8 cm

18. Pie charts

1. $\frac{72}{360} \times 2000 = 400$ kg

19. Drawing pie charts

1.

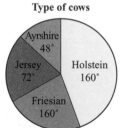

20. Bar charts

1. (a)

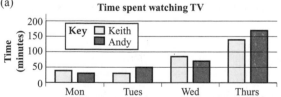

(b) The amount of time that Andy spent watching TV increased throughout the week.

21. Pie charts with percentages

1.

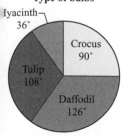

22. Using comparative pie charts

1. $\sqrt{\frac{450}{200}} \times 4 = 6$ cm

23. Frequency polygons

1.

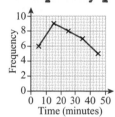

24. Cumulative frequency diagrams 1

1.

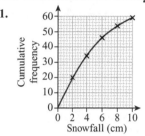

25. Cumulative frequency diagrams 2

1. (a) 44%   (b) 40 years of age

26. Histograms with equal intervals

1.

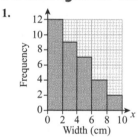

2. 35

27. Histograms with unequal intervals

1.

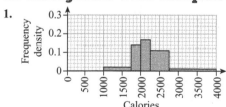

28. Interpreting histograms

1. (a)

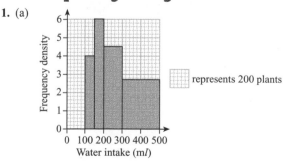

represents 200 plants

(b) 51%

ANSWERS

29. Population pyramids

1. The proportion in the oldest group for both men and women will have risen greatly by 2050 (about 15% in 2050 as against a very small percentage in 2000). The proportions of males and females in the youngest age group will have stayed roughly equal.

30. Choropleth maps

1.

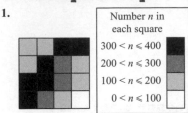

Number n in each square
$300 < n \leqslant 400$
$200 < n \leqslant 300$
$100 < n \leqslant 200$
$0 < n \leqslant 100$

2. The insects are distributed unevenly, with a band of higher insect density from bottom left to top right of the field.

PROCESSING DATA

31. Mode, median and mean

1. (a) (i) 18 (ii) 18 (iii) 20.3
(b) Less than the mean, as 20 is less than 20.3.

32. Mean from a frequency table

1. 2.4

33. Mean from a grouped frequency table

1. (a) 347 m² (3 s.f.)
(b) The data is grouped so the actual size of each plot of wasteland is unknown.

34. Median and mode from a frequency table

1. (a) 24 (b) 24

35. Averages from grouped frequency tables

1. (a) (i) $0 \leqslant D < 5$ (ii) $10 \leqslant D < 15$
(b) 10.9

36. Which average?

1. (a) 2 (b) 2.1 (1 d.p.)
(c) The mean will increase by 0.4. The median will not change.

37. Estimating the median

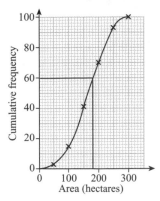

1. 190 hectares

38. The mean of combined samples

1. 6.7
2. 166

39. Weighted means

1. 70%

40. Measures of spread

1. $Q_1 = 1$, $Q_3 = 4$, IQR = 3

41. Box plots

1.

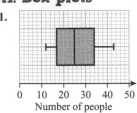

42. Interquartile range and continuous data

1.

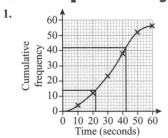

IQR = 20 seconds

43. Percentiles and deciles

1. (a)

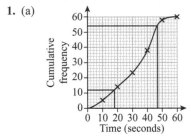

(b) (i) 18 seconds (ii) 47 seconds

44. Comparing discrete distributions

Group A: 0 1 2 3 4 5 6 6 7 8 8 9 10 12 13
Median = 6, $Q_1 = 3$, $Q_3 = 9$, IQR = 6
On average, group A went to fewer shops.
Group B visits are more clustered about the median.

45. Cumulative frequency diagrams and box plots

1.

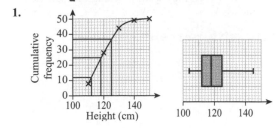

46. Using cumulative frequency diagrams and box plots

1.

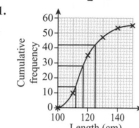

The median length of snakes in the zoo is 117 cm so on average they are shorter than snakes in the wild.
The interquartile range of snakes in the zoo is 13 cm so the central 50% of the distribution of their lengths is narrower than snakes in the wild.

47. Box plots and outliers

1. (a) $Q_1 = 8$, $Q_3 = 17$ (b) There are no outliers.

48. Box plots and skewness

1. (a)

(b) The median is almost exactly in between the two quartiles so the distribution is symmetrical. There is no tendency for either a large attendance or a small attendance.

49. Variance and standard deviation

Mean = 20
Variance = 80
Standard deviation = 8.9

50. Standard deviation from frequency tables

Mean = $\frac{1200}{30}$ = 40

Variance = $\frac{51000}{30}$ − 1600 = 1700 − 1600 = 100

Standard deviation = 10
(a) Standard deviation = 4
(b) Mean will stay the same, standard deviation will go down.

51. Simple index numbers

(a) 171.9   (b) £4.98

52. Chain base index numbers

(a) 102 and 102
(b) The chain base means that the base values for the two years are different.

53. Weighted index numbers

106.5

54. Standardised scores

Aptitude 0.3, skills 1.125
(a) −0.5   (b) 64

CORRELATION AND TIME SERIES

55. Scatter diagrams and correlation

1. (a) Explanatory – age, response – weight
Positive correlation
(b) Explanatory – length, response – time
Positive correlation
(c) Explanatory – temperature, response – number of cups of hot tea drunk
Negative correlation
(d) Neither, as there is unlikely to be any correlation

56. Lines of best fit

1. (a)

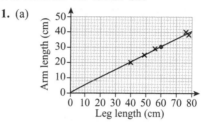

(b) 33 cm

57. Interpolation and extrapolation

1. (a) (i) £10500   (ii) Reliable as within the data
(b) (i) £1000   (ii) Unreliable as it is outside the data

58. The equation of the line of best fit

1. (a) $t = -1.4S + 12$
(b) The decrease in acceleration time for every litre increase in engine size.
(c) It predicts an acceleration time even when there is no engine present.

59. Curves of best fit

1. (a)

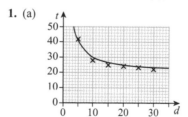

(b) 32.5 °C

60. Spearman's rank correlation coefficient 1

1. (a) −0.61
(b) The negative correlation coefficient indicates that the teams who score a large number of goals, also concede few goals, and vice versa.

61. Spearman's rank correlation coefficient 2

1. (a) 0.95
(b) There is a very strong positive correlation so the longer the bird the larger the weight.

ANSWERS

62. Time series

1. (a)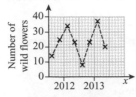

(b) Slightly downwards
(c) Quarter 3 of 2013

63. Moving averages

1. (a) 27.5 32.5 36 37.5 43 45 47 47.5
(b) The trend in the sales is increasing.

64. Moving averages and trend lines

1. (a)–(c)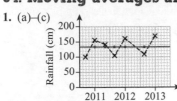

(d) The trend is flat – the rainfall is neither increasing nor decreasing on average.

65. The equation of the trend line

1. $y = -4x + 8276$ where x is the year and y is the number of moths

66. Prediction

1. (a)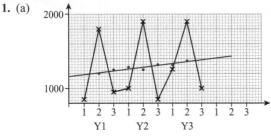

(b) ≈570   (c) ≈2000

PROBABILITY

67. Probability

1.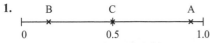

68. Sample spaces

1. (a) (1H, 2H), (1H, 2T), (1T, 2H), (1T, 2T) where (1H, 2H) means Head on the 1p coin and Head on the 2p coin.
(b) One head and one tail is the most likely
2. (a) (A, B), (A, C), (B, A), (B, C), (C, A), (C, B)
(b) A man and a woman is more likely.

69. Probability and sample spaces

1. $\frac{7}{10}$
2. (a) (1, 1) (1, 2) (1, 3) (1, 4)
(2, 1) (2, 2) (2, 3) (2, 4)
(3, 1) (3, 2) (3, 3) (3, 4)
(4, 1) (4, 2) (4, 3) (4, 4)
where (2, 1) means a 2 on the first spinner and a 1 on the second spinner.
(b) $\frac{2}{16}$

70. Venn diagrams and probability

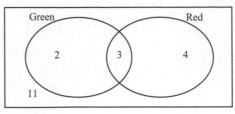

(b) $\frac{11}{20}$

71. Mutually exclusive events

1. (a) 0.28   (b) 0.4

72. Independent events

1. (a) 0.3   (b) 0.4 × 0.25 = 0.1
(c) 0.05 × 0.05 = 0.0025

73. Probabilities from tables

1. (a)

		Year group			
		Year 9	Year 10	Year 11	Total
Diet	Vegetarian	15	8	4	(27)
	Not vegetarian	19	22	7	48
	Total	(34)	(30)	11	75

(b) (i) Not vegetarian
(ii) $\frac{30}{75}$
(iii) $\frac{4}{75}$

74. Experimental probability

1. (a) $\frac{110}{250}$ (b) $\frac{140}{250}$
2. (a) Because he rolls the dice the most times.
(b) $\frac{36}{150} = 0.24$
(c) Combine all three people's results.
$\frac{75}{300} = 0.25$

75. Risk

1. (a) Football 0.16, hockey 0.083, rugby 0.325
(b) 2.9   (c) £33

76. Probability trees

1. (a)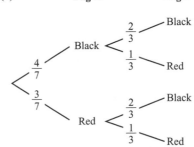

(b) $\frac{10}{21}$

77. Conditional probability

1. $\frac{42}{90}$

78. Probability formulae

(a)

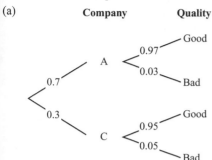

Company — Quality

A — 0.97 Good / 0.03 Bad
0.7
0.3
C — 0.95 Good / 0.05 Bad

(b) (i) 0.7 (ii) 0.97 (iii) 0.679 (iv) 0.964

79. Simulation

(a) Use 0, 1, 2 for 0 people arriving in any minute.
Use 3, 4, 5, 6, 7, 8 for 1 person arriving in any minute.
Use 9 for 2 people arriving in any minute.

(b)

Minute	1st	2nd	3rd	4th	5th	6th	7th	8th	9th	10th
Number joining	0	0	0	0	1	2	1	1	1	1

(c) 7 in the queue (assuming none in the queue to start with)

80. Probability distributions

1. (a) $\frac{3}{16}$ (b) $\frac{10}{16}$

81. The binomial distribution 1

1. (a) 0.32   (b) 0.64

82. The binomial distribution 2

1. (a) 0.0000759375   (b) 0.0021515625   (c) 0.026611875

83. The normal distribution

1. (a) 2.8 kg   (b) 1.8 kg to 3.0 kg

84. Quality control 1

1. (a) Warning limits 325.2 ml and 334.8 ml.
Action limits 322.8 ml and 337.2 ml
(b) The mean is 283 ml so the production line should be stopped.

85. Quality control 2

1. Warning limits 118 g and 132 g
Action limits 114.5 g and 135.5 g

For your own notes

For your own notes

Published by Pearson Education Limited, 80 Strand, London, WC2R 0RL.

www.pearsonschoolsandfecolleges.co.uk

Copies of official specifications for all Edexcel qualifications may be found on the website:
www.edexcel.com

Text © Pearson Education Limited 2015
Edited by Gordon Davies and Linnet Bruce
Typeset by Tek-Art, West Sussex
Original illustrations © Pearson Education 2015
Cover design by Miriam Sturdee

The right of Rob Summerson to be identified as author of this work has been asserted by him in
accordance with the Copyright, Designs and Patents Act 1988.

First published 2015

18 17 16 15
10 9 8 7 6 5 4 3 2 1

British Library Cataloguing in Publication Data
A catalogue record for this book is available from the British Library

ISBN 978 1 292098296

Printed in Slovakia by Neografia

Acknowledgements
All images © Pearson Education Limited

Every effort has been made to contact copyright holders of material in this book.
Any omissions will be rectified in subsequent printings if notice is given to the publishers.